# Post-Western Social Sciences and Global Knowledge



VOLUME 6

The titles published in this series are listed at *brill.com/psgk*

Disasters, Recovery and Emotions in Asia and in Europe

# Disasters, Recovery and Emotions in Asia and in Europe

*Edited by*

Laurence Roulleau-Berger
Yoshiyuki Yama

BRILL

ISSN 2352-5827
ISBN 978-90-04-75175-0
ISBN 978-90-04-75180-4 (PDF)
DOI https://doi.org/10.1163/9789004751804

Library of Congress Control Number: 2025052112



Brill and Koninklijke Brill BV are part of De Gruyter Brill.
www.degruyterbrill.com

Questions about General Product Safety Regulation: productsafety@degruyterbrill.com

Cover illustration: "*Tsunami*" painting by Francis Berger, 2022. Courtesy of the artist.

# Contents

# Figures and Tables

## Figures

## Tables

## Notes on Contributors

*Mara Benadusi*
is Professor of Anthropology at the University of Catania. Her work in disaster anthropology explores the social, political, and environmental dimensions of post-catastrophe reconstruction, ecological crises, and industrial risk. She has conducted long-term ethnographic fieldwork in Sri Lanka following the 2004 tsunami and in southeastern Sicily, where her work examines the lasting legacies of oil extraction and the complexities of energy transition in shaping local imaginaries and plans for recovery. She is co-founder of the Disaster and Crisis Anthropology Network within the European Association of Social Anthropologists and has published extensively in leading international journals and edited volumes. She is currently Co-Principal Investigator of the Marie Curie doctoral Network "C-URGE: The Anthropology of Global Climate Urgency".

*Jin Chen*
is an Associate professor in the School of Political Science and International Relations at Tongji University, Shanghai, China. His research interests include ritual discourse and action, the social and cultural memory of time and space, knowledge production, and the history of anthropology in Europe. He has published articles in Chinese, French and English in peer-reviewed journals, including *Minzu Yanjiu* ("Comprehensive practices and multiple relations: the "gubu" ritual of the Na's Daba"), *Temporalités* ("Modernité compressée, mémoires et lieux dans deux villes d'eau: Tongli et Zhenze"), and *The Journal of Chinese Sociology* ("The 'Body of ritual' transformed: discourses and actions in the Daba rituals among the Na (Moso)"). He has also translated into Chinese the works of French scholars such as Maurice Godelier and Lionel Obadia.

*Rita Ciccaglione*
holds a Ph.D. in Cultural Anthropology from the University of Rome La Sapienza and is currently a research fellow at the University of Naples L'Orientale. Her research sits at the intersection of disaster anthropology and urban anthropology, with a particular focus on how environmental risk and disaster management policies interact with broader processes of spatial reconfiguration, social transformation, and livelihood strategies in urban settings. She has conducted fieldwork in various Italian regions, including post-earthquake reconstruction sites and urban areas affected by hydrogeological instability and water scarcity.

Among her main investigations is a long-term research project in the aftermath of the 2009 earthquake in L'Aquila. Her work has been widely published in both national and international academic journals.

*Benoit Giry*

Ph.D. in sociology, is Associate Professor of sociology at Sciences Po Rennes (University of Rennes, France), where he co-leads the RESSP research program. His current research focuses on disaster governance and the effects of disasters on inequalities and economic, political, and social behavior. His publications include: *Sociologie des catastrophes* [Sociology of Disasters], La découverte, 2023; « Les désastres existent-ils? Exercice définitionnel préalable à toute sociologie des catastrophes future » [*Is There Such a Thing as a Disaster? An Examination of Some Definitional Issues*]. *L'Année Sociologique* 75 (2), 2025; « RestezChezVous » ou l'Etat sanitaire distribué. Sociologie des contributions numériques ordinaires à la mise en œuvre du confinement » *« RestezChezVous » or the Distributed Health State. Sociology of Ordinary Digital Contributions to the Implementation of the French Lockdown*]. *Gouvernement et action publique* 4 (12), 31–59, 2023.

*Koichi Hasegawa*

President at Morioka University and Professor Emeritus at Tohoku University. He received his Ph.D. from the University of Tokyo. His research interests are environmental sociology, civil society, social movements, and social change. His representative works include: *Constructing Civil Society in Japan: Voices of Environmental Movements* (Trans Pacific Press, 2004) and *Beyond Fukushima: Toward a Post-Nuclear Society*. (Trans Pacific Press, 2015), *Climate Change Governance in Asia* (Coeditor, Routledge, 2020), "Japanese Environmental Sociology: Focus and Issues in Three Stages of Development," (*International Sociology Reviews* Vol. 36, No.2, 2021), *Air Pollution Governance in East Asia* (Coeditor, Routledge, 2022).

*Bonjun Koo*

is a Researcher at the Disaster Reduction and Human Renovation Institution (DRI) in Kobe, Japan, a public centre established to preserve the lessons of the 1995 Great Hanshin-Awaji Earthquake and promote disaster risk reduction and community resilience. Previously, he was an Assistant Professor at the *Center for the Creation of Symbiosis Society with Risk* at Yokohama National University in Yokohama, where he conducted research on the social acceptance of energy systems and risk communication. Prior to that, he was a researcher at the Research Institute for Humanity and Nature in Kyoto, where he worked extensively on transdisciplinary approaches and stakeholder engagement methodologies. His specialisation lies in disaster risk management and risk communication, which he approaches through community participatory methods. He earned both his MS and PhD in Urban Management from Kyoto University.

*Thung-Hong Lin*

is Research Fellow, Institute of Sociology, Academia Sinica (2019–) Stanford-Taiwan Social Science Fellow, CASBS, Stanford University (2023–24). After receiving a Fulbright Scholar (2023–24) he has got his Ph.D. in Social Science, Hong Kong University of Science and Technology. His research focuses on class inequality, social mobility, and the political economy of Taiwan and China. He also examines the sociology of disaster and climate change, emphasizing how structural inequality shapes risk and resilience. He employs quantitative and comparative methods across East Asia and global datasets. Selected Publications: "Governing Natural Disasters: State Capacity, Democracy, and Human Vulnerability." *Social Forces*, 93(3) (2015); "How Does Resilience Matter? An Empirical Verification of the Relationships Between Resilience and Vulnerability." *Natural Hazards*, with E. Lin, H.C. Lee 88(2), 1229–1250, 2017– "Government-Sponsored Disinformation and the Severity of Respiratory Infection Epidemics Including COVID-19: A Global Analysis, 2001–2020." With M.C. Chang, C.C. Chang, Ya-H.Chou, *Social Science & Medicine*, 296, 114744, 2022– "For Security or Sustainability? Investigating the Global Nexus of Nuclear Power, Democracies, and Civil Society." With Y.C. Ting *Energy Research & Social Science*, 81, 102284, 2021–.

*Vincenzo Lo Bartolo*

graduated in History from the University of Pisa and is currently specializing in History and Cultures of the Mediterranean at the University of Catania. His anthropological research focuses on post-earthquake relocation processes and territorial transformations following seismic events in the Mount Etna region. He is a member of the research project "Spaesamento e appaesamento alle pendici del Monte Etna", conducted by the Department of Political and Social Sciences at the University of Catania in collaboration with the National Institute of Geophysics and Volcanology (INGV). In addition to his academic work, Lo Bartolo is actively involved as a volunteer with the Italian Civil Protection in Sicily.

*Pierre Manoury*

Ph.D. student in Sociology at Lyon 2 University (France) and Renmin University of China, Triangle CNRS research unit, under the co-supervision of Professor Laurence Roulleau-Berger, Research Director Emeritus at CNRS, Ecole Normale

Supérieure of Lyon, Triangle, and Zhao Yandong, Professor of Sociology at Renmin University, Beijing. His thesis focuses on the sociology of the environment, risks, and organizations, adopting a crossed perspective between China and France. The title of the dissertation is: *"For a Sociology of Suspicion: Environmental Crises, Emotions, and Mobilizations in China and France."*

*Mario Mattia*
is a geophysicist at the Etnean Observatory of the INGV (National Institute for Geophysics and Volcanology) in Italy. For over 30 years, he has worked on geodesy applied to the modelling of volcanic and seismic sources. He has coordinated monitoring activities during numerous eruptions of Mount Etna and Stromboli, and he currently oversees the INGV branch in Messina. In addition to his scientific work, he has also collaborated with scholars in the human sciences on post-earthquake relocation processes and the study of territorial evolution in the aftermath of seismic events. He has authored and co-authored numerous publications in national and international journals in the fields of geophysics, volcanology, and disaster studies, also in collaboration with social scientists.

*Laurence Roulleau-Berger*
is Research Director Emeritus at CNRS, Ph.D. Supervisor in Sociology, Triangle, École Normale Supérieure of Lyon, and was the French Director of the International Advanced Laboratory CNRS/ENS Lyon-Chinese Academy of Social Sciences *Post-Western Sociology in Europe and in China* (2013–2024). She has led numerous research programs in Europe and China in urban sociology, economic sociology, sociology of migration and sociology of disaster. Since 2006, she has opened an epistemological way in the fabric of Post-Western Sociology. She has published more than 30 books and 200 journal articles and chapters, among which: *Post-Western Revolution in Sociology. From China to Europe* (2016); *Ecological Risks and Disasters—New experiences in China and Europe* co-edited with Li Peilin; (2016) *Work and Migration. Chinese Youth in Shanghai and Paris*, co-authored with Yan Jun (2017); *Young Chinese Migrants, Compressed Individual and Global Condition* (2021); *Sociology of Migration and Post-Western Theory*, co-edited with Liu Yuzhao (2022); *Handbook of Post-Western Sociology. From East Asia to Europe*, co-edited with Li Peilin, Kim Seung-Kuk, and Shujiro Yazawa (2023). She has been appointed in 2025 Vice-President of the Global China Academy, British Academy. Among awards, in 2021, she received from the *French Medal of Chevalière de l'Ordre National du Mérite* for her scientific work.

*Oscar Truong*
obtained a doctorate in sociology in 2023, Ecole Normale Supérieure de Lyon (ENS Lyon). He is associate researcher in Triangle, CNRS/ENS Lyon. His research explores socially engaged art practices in China, focusing on the emergence of independent artists spaces in Guangzhou, Shanghai, and Wuhan. His research highlights the transnational networks of young people in place-making, activist exchanges from China to East Asia. He is based in Taipei, where he conducts postdoctoral research on transnational youth activism, the geopolitics of social movements, and the mapping of transcritical archipelagos.

*Yoshiyuki Yama*
Ph.D. in Sociology, is Professor and currently serves as Dean of the School of Human Welfare Studies and Director of the Institute of Disaster Area Revitalization, Regrowth and Governance at Kwansei Gakuin University in Japan. He has held adjunct or visiting professor positions at institutions including the Earthquake Research Institute of the University of Tokyo, the Disaster Prevention Research Institute of Kyoto University, the National Museum of Ethnology, the International Research Center for Japanese Studies, the University of International Relations Beijing, Beijing Foreign Studies University, and Yunnan Minzu University, and served as visiting researcher at Université Paris Diderot Paris 7. His research interests are sociological theory, sociology of disaster, and heritage studies, particularly on local development and community-based disaster risk management in depopulated areas of Japan. He has authored or edited more than thirty books and published numerous scholarly articles in sociology, anthropology, disaster studies, heritage studies, and the history of Japanese philosophy.

*Qiaoyun Zhang*
is an Associate Professor in the Cultural Heritage and Anthropology Ph.D. Program at Macao Polytechnic University, Macao, China. Her research focuses on disaster anthropology, intangible cultural heritage safeguarding, and digital society. Dr. Zhang has authored over ten articles in leading peer-reviewed journals and has secured research funding from multiple international and domestic agencies.

# Introduction: Sociology of Disaster and Post-Western Theory

*Laurence Roulleau-Berger and Yoshiyuki Yama*

In this book, we start from a disaster sociology that extends the thinking of Félix Guattari (2008) with his theory of the Three Ecologies, which is based on the interrelation of mental ecology, social ecology, and environmental ecology. This theory has inspired Isabelle Stengers (2019) and leads to the idea of a decolonial ecology. Thinking of a decolonial ecology means a critique of the hegemonic thinking of the Global North in order to move towards non-hegemonic and dialogical sociologies. We will offer a glimpse beyond the "East" and the "West" by opening the horizons to a wealth of self-directed narratives by societies worldwide, thus laying the foundations for a Post-Western space. This is a place where "Western" and "non-Western" sociologies can intersect and interact, forming shared divergent understandings of myriad *ethnoscapes* (Appadurai, 1996). From the production of an epistemology shared with European and Chinese sociologists for more than 15 years, and more recently with Japanese and Korean sociologists, a *Post-Western Sociology* (Roulleau-Berger, Li, Kim, Yazawa, 2023) enables a dialogue—held on an equal footing—on European and Asian theories, to consider the continuities and discontinuities, the conjunctions and disjunctions between knowledge spaces situated in different social contexts (Roulleau-Berger, 2021). From this *ecology of knowledge* we can observe, on the one hand, the multiplication of *autonomous knowledge* (Alatas, 2022) vis-à-vis Western hegemonies, and, on the another hand, epistemic assemblages between European and Asian sociologies. This groundbreaking contribution is to co-produce a Post-Western Space in a cross-pollination process where "Western" and "non-Western" knowledge about disaster do interact, articulated through cosmovisions, as well as the co-production of transnational fieldwork practices.

## 1 Towards Post-Western Theory

Since 1979, diverse Chinese sociology theories have been developed from different perspectives, attesting to a real internationalization of the discipline and the solidification of new boundaries. Chinese sociology gained epistemic autonomy before 1949 and since 1979 outside the dichotomy between the "West" and the

 HTTPS://DOI.ORG/10.1163/9789004751804-002

"non-West," the colonist and the colonized, outside post-colonial sociology (He Yijin, 2018). Sociology in China has been reconstructed around the increasing importance attached to the effects of Chinese civilization, both past and present, and the idea of producing paradigms free of any form of cultural hegemony (Li, Li, and Ma, 2008; Roulleau-Berger, Guo, Li, and Liu, 2008). Therefore, contemporary Chinese sociologies appear to stand within a mosaic of constructivisms against backgrounds of historical or civilizational contexts (Li and Qu, 2011; Xie, 2012; Roulleau-Berger, 2016). Li Peilin and Wei Jianwen (2018) considered "Post-Western Sociology" as neither a weapon to deconstruct the hegemony of Western discourse in the post-colonialist discourse nor the further "ideologicalization" in the concept of "East." Instead, they attempt to construct a sociological knowledge system beyond the binary opposition of the West and the non-West.

Then, starting from 2000, the major wave of the sinicization of sociology defined itself in contrast with hegemonic Western thought and indigenization in the international academic environment. Chinese sociologists agree with the idea that Western sociologies should not be considered antagonistic to Chinese sociology. Xie Lizhong (2021) presents a perspective by classifying the indigenous sociologies developed in a non-Western country like China into four possible types: "object-transformed indigenization," "supplemented-modified-renewed indigenization," "theoretical substitution indigenization," and "theoretical-methodical substitution indigenization."

In South Korea, Shin Kwang-Yeong (2013) identified three modes of hegemonic social sciences constructed in a double indigenization of social sciences: the development of paradigms in the West, and the dominance of located concepts and theories associated with institutional power. The third one is the contested hegemony that refers to unavailable alternative theories. Kim Seung Kuk (2014) introduced the idea of the invention of an "East Asianism" to propose the orientalization of a Westernized East Asia from hybridizations of "Western" and "non-Western" knowledge, and to move towards a cosmopolitan society by constructing transnational regional identities. Kim Seung Kuk, to break the continuing myth of Orientalism in the name of liberating Easternization or East Asianism, considered one round hybridization on the one hand through the Easternization of Westernized East Asia, and, on the other hand, a second round hybridization between an emerging non-Western West and an Easternizing East Asia.

For Shujiro Yazawa (2014), Western sociology was diffused in Japan in the context of a process of cultural translation adapted to the Japanese academic field. This author explained that after 1980, a post-modern Japanese sociology was developed, with the reappearance of forgotten pre-war authors, such as Yasuma Takada and Kensuke Suzuki, and an indigenous sociological theory began to form. For Daishiro Nomiya (2019), if Japanese sociology started from the importation of Western sociology, the incorporation of varied sociological tradition has laid the foundation for the development of Japanese sociology, and now from sociological imaginations and empirical knowledge it is providing multiple perspectives on a global scale.

Japanese sociologists, such as Keiichi Sakuta, Akira Kurihara, Kenji Kosaka, and Munesuke Mita, also have developed their own concepts to elucidate the structural and cultural features of Japanese society which cannot be explained by Western traditional sociological theory. In a perspective of hybridization Yoshiyuki Yama (2021) combines the Western sociological theory with the Japanese confucianist theory of ritual to produce a non-Western hybrid or sociology of narration and narrative.

After the Fukushima disaster, Saburo Horikawa (2014) proposed theoretical answers by re-interrogating the notion of "damage" by dealing with its subjective dimension and by inviting the development of a theory of the commons; Daisaku Yamamoto and Hiroyuki Torigoe (2023) developed the theory of life environmentalism by focusing on the existence, the experience, the ways of life of individuals in communities and common spaces. We can see how the Fukushima disaster has produced new areas of situated knowledge and forms of epistemic autonomy.

Drawing on European and Asian experiences, we will analyze how a post-Western space has come into being in which sociological knowledge on disaster is produced that is both specific and shared and in which theories are gathered based on very different histories and traditions.

In sociology of disaster, we will investigate what makes *autonomous knowledge* in some European and Asian studies. Then, we will analyze the circulations and assemblages in transnational spaces of knowledge about disasters between Europe and Asia; this will make it possible to show the junctions, the tensions, the (in)compatibilities between acquired knowledge, knowledge produced locally, and knowledge implemented in different contexts (Roulleau-Berger, 2023).

## 2 Sociology of Disaster, Autonomous Knowledge, and Societal Singularities

According to Syed Farid Alatas (2024) we consider a plurality of autonomous knowledge in Post-Western sociology:

- *foundational knowledge* refers to the foundations of societies and their cultures,

- *consolidative knowledge* refers to knowledge that consolidates foundations,
- *reactive knowledge* refers to knowledge required the basis of social life.

### 2.1 *Autonomous and Foundational Knowledge*

In China Fei Xiaotong contributed significantly to producing foundational knowledge by drawing on a Confucianist perspective and modern social science theory (Li Youmei, 2023).

Foundational knowledge was produced before 1949 and in the 19th century, Yan Fu and Liang Qichao made major contributions to the emergence to the emergence of Chinese sociology and indigenous knowledge (Li Peilin,2023). In Japan, pre-modern Confucian scholars of the early modern period contributed to producing autonomous and foundational knowledge: for example, the Confucian concept of "*michi*" (Yama Yoshiyuki, 2021).

In sociology of disaster, the first issue to examine is the concept of "disaster." Contemporary societies face large-scale natural disasters driven by climate change, global-scale environmental destruction, and infectious diseases that pose serious threats to humanity. These recurrent, destructive events, which can also spark renewal and reconstruction, expose societies to what the Japanese disaster researcher Norio Okada terms "Persistent Disruptive Stressors (PDSs)" (Okada, 2021). If we take a broad view of "disaster" as PDSs, social challenges, such as declining birth rates, aging populations, and depopulation, can also be included. This broad definition corresponds to Thung-Hong Lin's view in Chapter 5, which frames epidemics, wars, earthquakes, and typhoons as the "Four Horsemen of the Apocalypse" in Taiwanese history.

Additionally, when comparing disaster contexts across Western and non-Western societies, it is also important to pay attention to the words that describe "disaster." In East Asian countries that use Chinese characters, academic discourse mostly adopts Sino-Japanese or Sino-Korean terms, and in Korea, even native terms often have Chinese character origins. Since the Meiji period in Japan, Western concepts have been introduced and translated into Sino-Japanese terms, some of which subsequently spread to China and Korea as well. This has facilitated scholarly communication across East Asia, yet it can also lead to an unfounded assumption that these terms necessarily have the same meaning across countries and cultures, resulting in unintended misunderstandings.

A notable example is the word "disaster" itself. In China and Taiwan, as in Japan, the term "災害" is commonly used, but the word "災難" is also prevalent. In Korea, "災難" is the most common word, even appearing in legal language. This linguistic difference often means that what the Japanese understand as "災害" is understood as "災難" in Korea and vice versa. However, this equivalence can be misleading. In Japan, "災害" usually evokes the image of a natural disaster, whereas "災難" in Korean contexts is more strongly associated with accidents, such as building collapses, bridge failures, and subway fires. Moreover, in Japan, different terms distinguish people affected by disasters, "被災者", and those harmed by accidents, "被害者", while in Korea one broad category is often applied across different kinds of suffering. Even in the case of natural disasters, the term that literally means "those who have suffered a disaster", "被害者", can also be used for victims of large-scale accidents, making distinctions less clear.

This issue became especially evident during the COVID-19 pandemic. In Korea, the pandemic was treated as a disaster from the outset, allowing for prompt action by researchers and response institutions. By contrast, in Japan there was less recognition of the pandemic as a disaster, and as a result, disaster researchers and institutions responded more slowly. The differences in conceptualization shaped differences in perception, which in turn influenced research approaches and governmental responses.

More than eight decades ago, seismologist Torahiko Terada (1878–1935), in his essay "Sainan Zakkō (災難雑考)" (1935), noted that human responses could greatly reduce damage. Terada advocated viewing natural disasters and human-made accidents together under the broader concept of "sainan (災難)," suggesting that an expansive understanding of disasters enables one to recognize a greater range of misfortunes—from natural disasters and human-caused accidents to epidemics—and also appreciate the accumulated knowledge, skills, wisdom, and experience humanity has developed to face and overcome these hardships. In this sense, Okada's concept of "Persistent Disruptive Stressors (PDSs)" can be considered a modern version of the broad "sainan (災難)" concept.

By re-examining the histories of disaster response across Western and non-Western societies and the historically shaped thoughts, cultures, and social relations surrounding disasters as "practical wisdom for sustainability," we can more fully appreciate the accumulated insights that may help us address the PDSs that challenge contemporary society.

According to Yoshiyuki Yama in Chapter 1, the concept of *fukko* literally means "to be revived again", and *symbolic recovery* could be assimilated to foundational knowledge in the sense that it refers to the foundation of Japanese society and culture. This notion takes on a central status in the way of thinking about the processes of reconstruction, regeneration, and transformation of societies damaged by disasters. Furthermore, Yama discusses that since the Great Kantō Earthquake of 1923, the dominant model of recovery led by national and local governments has been top-down urban redevelopment.

In contrast, other ideas have emerged that emphasize a people-centered recovery that prioritizes life reconstruction, as well as "symbolic recovery," which acknowledges the emotional and symbolic recovery of disaster-affected communities. This diversity of thought about *fukko* reflects ongoing debates and evolving interpretations. Similarly, *Koichi Hasegawa* in Chapter 3 explores lessons from the Great East Japan Earthquake, paying particular attention to changes in people's emotions. Hasegawa highlights the complex emotional structure that coexists, including positive feelings of trust and gratitude toward the state alongside negative feelings of dissatisfaction, distrust, and alienation. Moreover, he critically examines how the state tends to record and commemorate the disaster as a "success story," marginalizing the lived experiences and suffering of those most affected. *Zhang Qiaoyun* in Chapter 8 is introducing the concept of "cultural reconstruction" that should consider the endogenous and compassionate rebirth of the affected subject's intertwined social relations, the "relationship reconstruction" which lies in kinship, neighborhood, ritualistic practices, and supernatural beliefs. According to this author, Japanese psychologists, such as Haruo Hayashi, and Chinese anthropologists, such as Zhang Xi and Li Yongxiang, refer to a "culture of disasters" that is built on shared values, norms, beliefs, and meanings. All of these perspectives are inscribed in the conception of Eastern culture of "living together" (*kyosei*) with nature, including people / differences and similarities between the East and the West (Miwako Hosoda, 2023).

### 2.2 *Autonomous, Consolidative Knowledge and "Iconic Concepts"*

Another form of autonomous knowledge is consolidative knowledge, which refers to knowledge that consolidates foundations via some iconic concepts which will be introduced in this section (Yang Dian, 2023).

For Japanese sociologists, when examining natural disasters from a sociological perspective, it is essential to begin by acknowledging that disasters occur in the interplay between humans and nature. A landslide occurring deep in an uninhabited mountain range is merely a natural phenomenon; it is not a disaster unless it causes damage to human life or livelihoods. Nature provides humanity with many benefits, but it also generates disasters. Whether in rural communities closely tied to their surrounding environment, or in cities sustained by natural resources, people cannot live without a relationship with nature. Throughout history, humans have developed knowledge, practices, and techniques that allow them to enjoy nature's blessings, mitigate the risks it poses, and reduce the damage it causes—all in the service of sustaining their way of life. These accumulated strategies and innovations for dealing with recurring disasters in particular places can be referred to collectively as "disaster culture." In other words, people have created and preserved practical wisdom for preventing harm, coping with disasters, and sustaining their communities.

Over long periods of history, people have also created and passed down lessons learned from disasters to ensure that future generations can be better prepared. Survivors of devastating events have sought to make sense of these painful experiences by transforming them into cautionary lessons. For example, *Yoshiyuki Yama* in Chapter 1 introduces the tradition of *saigai denshō* (oral histories of disasters) in Japan, which includes local legends, folktales, proverbs, paintings, stone monuments, and ritual practices commemorating past catastrophes. Alongside these narratives, tangible traces of past disasters—preserved as disaster relics, memorial stones, and commemorative monuments—serve as "disaster heritage" or "disaster remains," functioning as social memory devices that transmit knowledge of past catastrophes across generations. *Disaster culture, disaster transmission*, and *disaster heritage* not only help preserve the memory of past disasters but also help communities heal from emotional trauma. They can thus be considered an important form of autonomous knowledge that disaster sociology must highlight.

In Japanese sociology, *disaster-related deaths* (*saigai kanrensi*), which refers to stress, deteriorating health, and social isolation as introduced by *Yoshiyuki Yama* in Chapter 1, appears as an iconic concept. In Chinese anthropology, *Jin Chen* in Chapter 10 has produced an approach focused on natural disasters, ethnic minorities' cosmology, and ecosystems. It offers an alternative perspective to the scientific understanding of natural disasters in their symbolic, moral, spiritual, and ecological dimensions. In the community indigenous knowledge represents a valuable resource for enhancing resilience, rituals hold pivotal significance in responses to disasters and we can understand how cultural practices intersect with environmental hazards and social vulnerabilities.

In Taiwanese sociology *Thung-Hong Lin* in Chapter 5, from the dynamic between state capacity and democratic civil society, refers to the concept of *institutional resilience* through the experiences of typhoons, earthquakes, SARS and COVID-19 and the risk of war between China and Taiwan. He considers the intersectionality of risks through the convergence of natural and geopolitical vulnerabilities in typhoons, leading to floods and landslides.

According to *Laurence Roulleau-Berger* in Chapter 2, in Western sociology, particularly in the French tradition, especially inspired by Pierre Bourdieu's theory, a sociology of inequalities has been produced, iconic concepts of environmental justice and of Capitalocène are used to demonstrate how domination, colonialism, and racism could produce ecological disasters. The author will explain how the disaster studies have provoked the rise of the sociology of emotions.

For *Benoît Giry* in Chapter 4 the fluidification of social institutions gives rise to anthropological figure: *homo emotionalis*, who is involved in an emotional economy where individuals and destiny communities are facing disasters. The *homo emotionalis* could feel fear, anxiety, despair, and anger, undergo the "disaster syndrome". The collective and individual capabilities to contribute to defining a "structure of feelings"—in the sense of Raymond Williams—in different societies and produce situated emotional socialization. The sociology of disasters invites us to integrate the *homo emotionalis* in a theory of action.

*Laurence Roulleau-Berger* and *Oscar Truong* in Chapter 11 consider that moral economies produce regimes of equity, peace, and abandonment of victims who are producing *emotional regimes* of consent, complaint, indignation, or distancing from the institutional responses in a post-disaster context. The interactions between moral economies and regimes of action are constructed around the figures of conciliation, negotiation, dispute, and exit between institutional actors and victims.

*Mara Benadusi, Rita Ciccaglione, Mario Mattia*, and *Vincenzo Lo Bartolo* in Chapter 6 have described the condition of disorientation associated with relocation through the experiences of *spaesamento* and *appaesamento*. In their anthropological approach they embrace cognitive, emotional, and sociocultural responses to disasters to explain how affected communities perceive, experience, and navigate risk, resilience, and recovery processes in experiential and affective stratigraphies. Here, the really innovative concepts like *spaesamento/appaesamento* and experiential and *affective stratigraphies* appear as iconic concepts and ones which produce autonomous knowledge.

### 2.3 *Autonomous, Reactive Knowledge and Dissonances*

Autonomous/reactive knowledge allows for the examination of the instances of *dissonances* between Western ethos and non-Western ethos based on unique experiences (Daishiro Nomiya, 2023). For example, in Japanese sociology the concept of *collective memories of trauma* developed by *Yoshiyuki Yama* in Chapter 1 or the *hibakusha collective memory* by Daishiro Nomiya in postwar history are translating dissonances between European and Japanese sociologies. According to *Zhang Qiaoyun* in Chapter 8, by showing the importance of cultural protection for economic development, Chinese disaster anthropology transcends Western critical theories of capitalism and colonialism to produce autonomous knowledge.

*Pierre Manoury* in Chapter 9 is focused on the fabric of suspicion in slow-onset disasters and he distinguishes the concept of *eco-anxiety* of individuals mobilized in ecological movements in democracy and *wuli* of individuals facing disasters under authoritarian regimes to qualify individual responses and the organization of collective mobilizations. *Mara Benadusi, Rita Ciccaglione, Mario Mattia*, and *Vincenzo Lo Bartolo* in Chapter 6 introduce the notion of *seismophobia* to describe this intense and persistent anxiety triggered by seismic tremors produced in a paradoxical process of "scotomization."

## 3 From East Asia to Europe: What Makes Common Knowledge about Disasters?

### 3.1 *Community, Emotions, and Memory*

Disasters not only cause the destruction of society but also become a catalyst for its reorganization, as highlighted by Norio Okada's concept of "Persistent Disruptive Stressors (PDSs)." However, what truly drives this reorganization is not the disaster itself, but rather the movements of people's wounded emotions and the actions that emerge from those emotions. In Chapter 4 *Benoît Giry* emphasizes that emotions are an indispensable analytical axis in the sociology of disaster. In this view, alongside the familiar *homo economicus* and *homo sociologicus*, there exists the implicitly assumed the figure of *homo emotionalis* representing human beings who exhibit distinct emotional behaviors in the face of disaster. Giry's observation is significant because it foregrounds the role of emotions as a dynamic driving force for social reorganization. Disaster sociology must therefore stand with survivors and pay close attention to the voices of their emotions, especially recognizing that these voices have often been marginalized and suppressed during the recovery process. Evaluating and making visible the work that emotions do in reshaping society is an important task for disaster sociology.

In Japanese and Chinese disaster sociology scholars are using the concept of *community* in distinguishing different forms. This concept becomes polyphonic depending on the types of disasters, societies, and cultures. For example, *Jin Chen* in Chapter 10 is introducing the community of ethnic minorities in China and in Japan. *Bonjun Koo* in Chapter 7 is exploring the question of the power of community agency, which affirms that local knowledge, mutual aid, and relational networks are foundational; he is exploring how community structures, memory institutions, and inclusive agenda are producing technical and moral resilience, both historical and visionary, how disaster governance prioritizes empathy, memory, and collective imagination. A resilient community relies on "self-help" (jijo), "mutual help" (kyojo), and "public help" (kojo), notions very close to *Laurence Roulleau-Berger* and *Oscar Truong's* perspective in Chapter 11, who are using the concepts of *geography of care* and *emotional community of destiny*, which produce new social solidarities, inequalities, and new moral boundaries.

For *Zhang Qiaoyun* in Chapter 8, honoring the memory of heroic martyrs and the daily engagements and narratives of the invisible lives helps us to understand how individuals and communities do accept their losses and move forward with their lives. This work of memory involves returning to the sites of the disaster and the development of tourism activities which promote the process of moral and psychological repair.

### 3.2 *Institutions, Resilience, and Collective Action*

In the French tradition, there are the sociology of "new social movement" by Alain Touraine, the dispositionalist sociology by Pierre Bourdieu, interactionist and pragmatist theories by Daniel Cefaï (2007).

In Japanese sociology there is the reinterpretation of the conception of a "risk society" with environmental and collective action: the social mechanism of suffering victims, recovery process, environmental justice, and social disparity (Hasegawa, 2023).

In China sociologists are using the notions of ecological civilization, social governance, social innovation, and social construction (Wang Xiaoyi, Asnier, 2023). The control of collective actions in the technocratization of an authoritarian political regime is analyzed from the American theories of collective action and the notion of "mass incident" (*quntixing shijian*) through the politization of social movements in the past and in the present (Feng Shizeng, 2023).

In different chapters we can understand how resilience has become a central category in disaster policies. For *Koichi Hasegawa* in Chapter 3, new problems emerge one after another in the reconstruction disaster process, and the reconstruction disaster itself can bring about "new disasters"; this author considers the nature of decision-making and policy-making, the merits and demerits of the reconstruction system; and for him the recovery process includes supports for disaster victims, recovery of homes and towns, and the recovery of industry and livelihoods.

Emotional recovery and politics of memory are associated by different categories of institutional and political actors, the victims in the reconstruction of the local society. Bonjun Koo in Chapter 7 defines trauma, grief, and memory as constituents of disaster governance.

In Japanese, Chinese, and European disaster sociologies public help and public arenas raise the questions of trust, mistrust, and suspicion at the center. *Pierre Manoury* in Chapter 9 considers the entrepreneurs produce different forms of suspicion; eco-anxiety and *wuli* emerge as two situated forms of suspicion provoked by slow-onset disaster in the process of institutional resilience.

The cross-analysis between Japanese sociology and French sociology has produced common knowledge, but we can speak of theoretical dissonances in sociology of collective action between French/Japanese sociology and Chinese sociology.

### 3.3 *Recovery, Time, and Space*

Before and after the disaster, as written by *Mara Benadusi, Rita Ciccaglione, Mario Mattia*, and *Vincenzo Lo Bartolo* in Chapter 6 the recovery process holds absence and presence, memory and forgetting, and past and future are held in a fragile and often contested balance. For *Zhang Qiaoyun* in Chapter 8 individuals are creating multiple meanings of the disaster in reconnecting with the past and the reconstruction of everyday life. This approach appears really connected with Yoshiyuki Yama and Jin Chen's perspective in Chapter 10; these theoretical assemblages are producing universal and situated knowledge.

*Laurence Roulleau-Berger* and *Yoshiyuki Yama* both examine how intermediate societal levels in Japanese and French societies through disaster studies and the development of the concepts of *intermediate spaces* and *intermediate intellectuals*, reconstructs individual and collective identities, and more broadly how societies "in disaster" work on themselves.

The sociology of emotions creates assemblages across Japanese, Chinese, and European perspectives focusing on emotional healing through a process of integrating and accepting traumatic experiences and strong emotions that generate the feelings of fear, rejection, and attachment. The status of emotions in international sociology is increasingly affirmed and disaster studies invites us to rethink the reconstruction of collective and individual identities, the diversity of capabilities, the resources repertoires, and the social and cultural inequalities.

## Bibliography

Alatas, S.F. 2022. Knowledge hegemonies and autonomous knowledge. *Third World Quarterly*, 1–18. https://doi.org/10.1080/01436597.2022.2124155.

Alatas, S.F. 2024. The Coloniality of knowledge and the Autonomous Knowledge Tradition, *Sociology Compass*, 18:e13256, https://doi.org/110.1111/soc4.13256.

Appadurai, A. 1996. *Modernity At Large: Cultural Dimensions of Globalization*. Minneapolis: University of Minnesota Press.

Cefaï, D. 2007. *Pourquoi se mobilise-t-on? Les théories de l'action collective*, Paris: La Découverte.

Feng, S. 2023. States's Temperament and the Control of Collective action in Contemporary China in Roulleau-Berger L., Li P., Kim S.K., Yazawa S., 2023. *Handbook of Post-Western Sociology. From East Asia to Europe*, Leiden&Boston, Brill Publishers.

Guattari, F. 2008. *Les Trois écologies*, Paris: Galilée.

Hasegawa, K. 2023. Contributions of Japanese Environmental Sociology in Non-Western context, in Roulleau-Berger L., Li P., Kim S.K., Yazawa S., 2023. *Handbook of Post-Western Sociology. From East Asia to Europe*, Leiden&Boston, Brill Publishers pp. 691–710.

He, Y. 2018. An Alternative Autonomy: The Self-adaptations of Chinese Sociology in the 1950s. In *Post-Western Sociology: from China to Europe*, ed. Roulleau-Berger, L., Li Peilin, Oxon/New York: Routledge, pp. 299–315.

Horikawa, S. (2014), Chronology as a Method: A New Challenge for Environmental Sociology in the Post-Fukushima Era. Sustainability and Environmental Sociology: Proceedings (Tokyo: the Institute for Sustainability Research, Hosei University), 297–308, Jul 13.

Hosoda, M. 2023. East-West Dialogue for Global Health Care Challenges in the Era of Covid 19 and Beyond in Roulleau-Berge L., Li P., Kim S.K., Yazawa S., 2023. *Handbook of Post-Western Sociology. From East Asia to Europe*, Leiden&Boston, Brill Publishers, pp. 928–937.

Kim, Seung Kuk, Li Peilin, Yazawa Shujiro, (eds) 2014. *A quest of East Asian Sociologies*, Seoul: Seoul National University Press, pp. 313–328.

Li, P., 2023. The Emergence and Characteristics of Chinese Sociology in Roulleau-Berger L., Li P., Kim S.K., Yazawa S., 2023. *Handbook of Post-Western Sociology. From East Asia to Europe*, Leiden&Boston, Brill Publishers, pp. 48–61.

Li, P., Li, Q., Ma, R., 2008. *Shehuixue he zhongguo shehui* (*Sociology and Chinese society*). Beijing: Shehui kexue wenxian chubanshe.

Li, P., Qu, J., 2011. *History of Sociology in China in the First Half of the Twentieth Century*. Beijing: Social Sciences Academic Press.

Li, P. Wei, J., 2018, Post-Western Sociology and contemporary Chinese sociology, in Roulleau-Berger, L., Li, P. (eds), Post-Western Sociology-From China to Europe, London and New York, Routledge Publishers, pp. 39–53.

Li, Y. 2023. Chinese Sociology: Traditions and Dialogues-Located knowledge Production as Post-Western Sociology in Roulleau-Berger, L., Li, P., Kim, S.K., Yazawa, S., 2023. *Handbook of Post-Western Sociology. From East Asia to Europe*, Leiden&Boston, Brill Publishers, pp. 143–157.

Nomiya, D. 2019. "A Short History of Japanese Sociology: Its Historical Legacies and Future Dreams." *Economic and Social Changes: Facts, Trends, Forecast* 12 (5): 155–157.

Nomiya, D. 2019. Voice of the Dead: Hibakusha Collective Memory against the Western Ethos, in Roulleau-Berger L., Li P., Kim S.K., Yazawa S., 2023. *Handbook of Post-Western Sociology. From East Asia to Europe*, Leiden&Boston, Brill Publishers, pp. 292–306.

Okada, N., 2021. "Zokuhatsu suru sainan dainamikusu no jidai to jizokukanō na chiiki fukkō e mukete" (続発する災難ダイナミクスの時代と持続可能な地域復興へむけて) [The era of recurring disaster dynamics and sustainable regional reconstruction]. *Saigai Kagaku* (自然災害科学), 40(3): 269–273. (in Japanese).

Roulleau-Berger, L. 2016. *Post-Western Revolution in Sociology. From China to Europe*, Leiden&Boston, Brill Publishers.

Roulleau-Berger, L. 2021. The fabric of Post-Western sociology: ecologies of knowledge beyond the "East" and the "West", *The journal of Chinese sociology*, April 2021, vol. 8, n° 10.

Roulleau-Berger, L. 2023. Toward Post-Western Sociology in Roulleau-Berger L., Li P., Kim S.K., Yazawa S., 2023. *Handbook of Post-Western Sociology. From East Asia to Europe*, Leiden&Boston, Brill Publishers, pp. 21–48.

Roulleau-Berger, L., Guo Y., Li P., Liu, S. (eds), 2008. *La nouvelle sociologie chinoise*, Paris: CNRS Publishers.

Roulleau-Berger L., Li P., Kim S.K., Yazawa S., 2023. *Handbook of Post-Western Sociology. From East Asia to Europe*, Leiden&Boston, Brill Publishers.

Shin, K.Y, 2013. «The emergence of hegemonic social sciences and strategies of non (counter) hegemonic social sciences», in Michael Kuhn et Shujiro Yazawa (dir.), *Theories About Strategies Against Hegemonic Social Sciences*, Tokyo, Center for Glocal Studies, Seijo University, p. 77–94.

Stengers, I. 2019. *Résister au désastre*, Paris: Wildproject.

Terada, T., 1935. "Sainan zakkō" (災難雑考) [Miscellaneous thoughts on disasters]. In Tetsuo Yamaori (ed.), *Tensai to Nihonjin: Terada Torahiko zuisaku-sen* (天災と日本人 寺田寅彦随筆選). Kadokawa Sophia Bunko, 2011. (in Japanese).

Wang, X., Anier, 2023. How Ecological Civilization Contributes to Post-Western Sociology in Roulleau-Berger L., L P., Kim, S.K, Yazawa S., 2023. *Handbook of Post-Western Sociology. From East Asia to Europe*, Leiden&Boston, Brill Publishers, pp. 779–786.

Xie Lizhong, 2021. Post-Western Sociologies: what and why? *The journal of Chinese sociology*, April 2021, vol. 8, n° 10.

Xie, L., 2012. Postsociology. Beijing: Social Sciences Academic Press.

Yama, Y., Sociology of ritual and narrative as post-Western sociology: from the perspective of Confucianism and Nativism in the Edo period of Japan. *The journal of Chinese sociology*, April 2021, vol. 8, n° 10.

Yamamoto, D., Torigoe, H., 2023. *Everyday Life-Environmentalism. Community sustainability and resilience in Asia*, London: Routledge.

Yang, D. 2023. Chinese Economic Sociology: From the perspective of Post-Western Sociology in Roulleau-Berger L., Li P., Kim, S.K. Yazawa, S., 2023. *Handbook of Post-Western Sociology. From East Asia to Europe*, Leiden&Boston, Brill Publishers, pp. 251–274.

Yazawa, S. 2014. Civilizational Encounter, Cultural Translation and Social Reflexivity: a note on History of Sociology in Japan. In, A quest of East Asian Sociologies, ed. Kim, S.K., Li P., Yazawa, S., 131–167. Seoul: Seoul National University Press.

CHAPTER 1

# Disaster Recovery in Japan: the Concept of "Fukkou," Memory, and Emotional Recovery

*Yoshiyuki Yama*

## 1 "Fukkou": the Concept of Disaster Recovery in Japan

To sociologically study disaster recovery in Japan, one must first understand the uniquely developed concept of *fukkou* (復興). As described below, although the term *fukkou*, written in Chinese characters, literally means "to be revived again," its interpretation varies depending on perspective even within Japan, leading to complex disputes.

Due to the frequent occurrence of natural disasters—such as earthquakes, tsunamis, typhoons, and torrential rains—Japan has developed advanced science and technology related to disaster prevention (*bousai*, 防災), including research in the natural sciences on earthquake mechanisms and seismic reinforcement of buildings. Similarly, post-disaster responses have focused on engineering studies of physical recovery (*fukkyuu*, 復旧) of infrastructure like roads and water supply systems.

In Japan, *fukkou* has traditionally been perceived as urban redevelopment through top-down public works led by the national and local governments. This view dates back to the 1923 Great Kanto Earthquake, when the slogan "Imperial Capital Reconstruction" (*Teito Fukkou*) framed the disaster as an opportunity to rebuild Tokyo as a capital befitting the Empire of Japan. This marked the beginning of the now-prevalent top-down, government-led concept of *fukkou* as urban redevelopment. A similar notion reappeared in 1995 following the Great Hanshin-Awaji Earthquake with the idea of "Creative Reconstruction" (*Souzouteki Fukkou*).

When disaster-hit cities are redeveloped, new buildings rise and streets are beautified, creating the appearance of progress. However, rising land prices often force long-time residents to leave, and reconstruction projects may ironically inflict further harm on survivors—a phenomenon known as "reconstruction-induced disaster." Architect Yoshimitsu Shiozaki, who has studied post-earthquake housing reconstruction and community planning, coined the term "Reconstruction Disaster" (*fukkou saigai*) to describe this dynamic (Shiozaki 2014).

HTTPS://DOI.ORG/10.1163/9789004751804-003

One well-known example of a reconstruction disaster is the forced redevelopment around Shin-Nagata Station after the Great Hanshin-Awaji Earthquake, carried out under the banner of "Creative Reconstruction" without resident consent. The area was transformed into a modern cityscape with high-rise buildings, but rising land values displaced former residents and destroyed longstanding community ties. Many newly built spaces remain vacant, failing to revitalize the area while leaving the local government in financial deficit. Not only were survivors' voices excluded from the process, but the reconstruction itself exacerbated their suffering.

This top-down approach was echoed after the 2011 Great East Japan Earthquake. The Reconstruction Design Council's mission statement emphasized that "we must pursue creative reconstruction aimed at the future, not merely restore the past." Similarly, the "Expert Council on Kumamoto Earthquake Recovery" presented proposals in 2016 for "Creative Reconstruction" following the 2016 Kumamoto Earthquake. In Mashiki Town, one of the worst-affected areas, roads were widened to four lanes in the name of reconstruction, resulting in the demolition of homes built on land now designated for expanded roads.

However, the Great Hanshin-Awaji Earthquake, which struck on January 17, 1995 and resulted in over 6,400 deaths, also revealed the limitations of conventional urban redevelopment models. As Japan entered a mature society following decades of rapid economic growth, this urban disaster struck a major city—Kobe—highlighting new vulnerabilities.

One key issue exposed was *disaster-related deaths* (*saigai kanrenshi*), which refer to fatalities not directly caused by the disaster itself, but occurring later due to stress, deteriorating health in shelters, or social isolation. These are especially common in earthquake disasters and are termed *shinsai kanrenshi*. Out of approximately 6,400 deaths from the Hanshin-Awaji Earthquake, around 940 were disaster-related. In the 2011 Great East Japan Earthquake, about 3,700 out of 22,000 deaths or missing persons were disaster-related. The 2016 Kumamoto Earthquake saw 218 disaster-related deaths out of 273 total—meaning such deaths comprised the overwhelming majority (Reconstruction Agency 2012; Hyogo Prefecture 2005; Kumamoto Prefecture 2018).

Additionally, more than 43% of the deceased in the Hanshin-Awaji Earthquake were aged 65 or older, meaning elderly individuals—disaster-vulnerable people—died at about three times the population ratio. This highlighted issues such as evacuation difficulties for the elderly, stress, isolation, and the importance of maintaining social connections and community. The need for *mental health care* (*kokoro no kea*) also came to prominence during this disaster.

Japan's legal framework to support disaster survivors remains underdeveloped. The Disaster Relief Act (1947) provides aid only immediately after a disaster and primarily in the form of material support. Temporary housing is limited to two years. While the Act on Provision of Disaster Condolence Grant (1973) exists, it is insufficient; once victims lose homes or property, escaping poverty becomes difficult. The support system is not survivor-centered.

The Act on Support for Reconstructing Livelihoods of Disaster Victims (1998), established after the Hanshin-Awaji Earthquake, offers housing aid based on damage assessments. Yet even the maximum grant of 3 million yen is often inadequate, and discrepancies in eligibility breed inequality and social division. Reconstruction efforts can even destroy the social ties essential to daily life.

As a result, many survivors resent the term *fukkou*. While it may sound positive, in reality it often refers to a form of recovery that excludes the voices of survivors. The same problems repeat with every disaster. This is because the most severely affected areas tend to be socially vulnerable or impoverished communities. In other words, survivors were already in precarious positions before disaster struck.

Thus, in contemporary Japanese society, the traditional top-down *urban redevelopment*-based understanding of *fukkou* is being questioned. Instead, new discussions focus on how to *rebuild* the lives of survivors and *revitalize* affected communities—a model more suited to a mature society, expressed through the evolving concept of *fukkou* as both "reconstruction" and "revitalization."

For example, Kojiro Miyahara, recognizing the tension between the conventional urban redevelopment model of "Imperial Capital Reconstruction" and post-Hanshin trends emphasizing "rebuilding lives," analyzed the inconsistent usage of *fukkou* across legal and administrative contexts. He advocated for a new "regenerative" model of recovery appropriate to a mature society (Miyahara 2005). This idea, however, has deeper roots. In response to the "Imperial Capital Reconstruction" rhetoric following the Kanto Earthquake, economist Tokuzo Fukuda (1874–1930) had already proposed a human-centered model of recovery (*ningen no fukkou*) (Fukuda 2012). He stated:

> I insist that the primary focus of reconstruction should be the recovery of human beings. By this, I mean the restoration of the opportunities for living that were destroyed by the disaster.

He further explained:

> Today, people must live, work, and conduct business in order to survive. Therefore, restoring opportunities for survival means restoring the opportunities for living, working, and doing business—collectively referred to as *eisei* (livelihood). Roads and buildings are merely tools to

> maintain and protect these livelihood opportunities. Even if roads and buildings are restored, it is meaningless unless the essential opportunities for livelihood are restored.

Shigeki Yamanaka has re-evaluated the history of this concept and discussed pathways to institutionalize such a philosophy (Yamanaka 2023). Yoshiteru Murosaki conducted a comprehensive analysis of Japan's disaster history, systematized the universal principles underpinning recovery, and underscored the necessity of addressing social distortions revealed by disasters, alongside the imperative of institutional preparedness and the allocation of human and financial resources prior to disaster onset (Murosaki 2020). Shingo Nagamatsu, synthesizing the work of the Japan Society for Disaster Recovery and Revitalization's committee "What Is *Fukkou*?" (2009–2011), categorized the concept into four approaches: (1) Ideological—what kind of society should be built; (2) Mechanistic—what elements are needed to achieve recovery; (3) Governance—how decisions are made and implemented; and (4) Capacity—what knowledge and skills are required. He also highlighted the political nature of the term *fukkou* (Nagamatsu 2020).

## 2 Symbolic Recovery

Amid the exploration of new concepts of *fukkou*, attention has been drawn to the idea of "symbolic recovery (*shouchouteki fukkou*)," proposed by the author from an anthropological perspective. In many disaster-affected regions, traditional festivals that serve as emotional anchors for local communities are suspended, and cultural heritage such as shrines, temples, and traditional buildings are destroyed—damaging the "symbols" of the community. The psychological wounds this inflicts on local residents are significant. It is argued that unless the people of the affected area themselves feel, "Now we have recovered," true recovery cannot be said to have been achieved. Based on this understanding, it emphasizes the importance of restoring local symbols as a vital part of the recovery process.

Furthermore, the author points out the importance of the officials responsible for recovery efforts appropriately designing and implementing "recovery rituals"—events that enable people to feel that recovery has taken place—and doing so at the right time. In this way, the author presents a new concept of recovery, distinct from traditional civil engineering- or urban redevelopment-based models, which focuses instead on the emotional healing, restoration, and revitalization of the people in affected areas. This is what the author calls "symbolic recovery."

The concept of "symbolic recovery" should now be explained in more detail. When people come to feel, "Now we have truly recovered," this realization must be understood as occurring at a level different from the usual physical and objective standards, such as the restoration of roads or infrastructure. At this level, a solution that is convincing to people can be described as a highly ritualistic act. If recovery itself is achieved through such ritualistic acts, it should be regarded not as a simplistic physical concept, but rather as a symbolic one.

If recovery is a symbolic concept, then we must actively utilize insights from fields like anthropology and sociology that examine the structure of symbols and the mechanisms of their formation. By doing so, we can identify methods for generating a shared sense of recovery among people.

Focusing on rituals that generate a sense of recovery among people is especially important because the community itself is symbolically constructed. This perspective, which has developed significantly in anthropology since Victor Turner's theory of ritual (Turner, 1969/1996), is now more broadly accepted. The idea that communities are culturally constructed is exemplified by works like Anthony Cohen's theory of the "symbolic community" (Cohen, 1985/2005) and Benedict Anderson's concept of "imagined communities" (Anderson, 1983/1997). Given that communities are symbolically or culturally constructed, the recovery of disaster-stricken communities must also be achieved at the symbolic level of people's meaning systems.

Particularly notable here is the work of Clifford Geertz. His theory of the "theatre state," which describes the former Balinese state of Negara as being realized through rituals—where the rituals themselves constituted the state—marked a significant shift in the understanding of ritual (Geertz, 1980/1990). He argued that "reality is as imagined as the unreal" (Geertz, 1980/1990:162). While the term "realistic" is typically used in opposition to the symbolic, this distinction is essentially superficial. In fact, what is considered realistic is often more imagined than the symbolic.

For instance, if roads and houses are material resources belonging to the realm of reality, then religious buildings and festivals are symbolic cultural resources. However, both types must be understood under the broader concept of the symbolic. This form of symbolism is realized within the community itself. Thus, for a community to recover, both physical resources (like roads and housing) and cultural ones (like religious buildings and festivals) must be restored. Alongside these recovery processes, a symbolic-level recovery must also be achieved. This is what we call "symbolic recovery."

However, symbolic recovery, achieved at the level of symbolic meaning systems, cannot be directly verified by visible means. Therefore, regardless of whether objective criteria—like infrastructure restoration—have been met, in order for people to acquire a sense of recovery, it becomes necessary to symbolically express recovery in a visible manner. Effective methods for expressing recovery include the restoration of symbols of the affected community, the resumption of local festivals, the repair or reconstruction of cultural heritage, the construction of disaster memorial monuments, and the organization of various events and festivals. These acts of symbolically expressing recovery will be referred to here as "recovery rituals."

The "recovery rituals" discussed here may correspond to what Émile Durkheim classified as "mimetic rites" (Durkheim, 1912/1975:210–242). Durkheim evaluated James Frazer's principles of sympathetic magic—"homeopathic magic" and "contagious magic"—favoring the former. According to Durkheim, mimetic rites provide only an image of the totem animal being abundantly born, with no preexisting model, and thus have inherent creative power (Durkheim, 1912/1975:220–221).

Drawing from Durkheim's insight, recovery, too, lacks a fixed model that can be pre-acknowledged as legitimate. What is provided is only the symbolic expression of the image of recovery. Hence, the recovery created through recovery rituals, as a symbolic image, constitutes what is called "symbolic recovery."

These recovery rituals can be said to have both manifest and latent functions in relation to objective standards. Manifestly, they function to certify that recovery has been achieved. When recovery rituals are performed, people judge that recovery has taken place. This manifest function typically operates when recovery is judged to have occurred in civil engineering, economic, or demographic terms—i.e., according to objective standards. Let us call this the "recovery recognition function."

Apart from this, there is also a latent function: the "recovery creation function," whereby recovery rituals generate recovery itself. Even if recovery is not recognized under objective standards, it is created at the symbolic level. Although recovery rituals appear to operate only via the recovery recognition function, the more crucial function is the recovery creation function. This is because the former is effective only when the latter is presupposed.

In summary, symbolic recovery refers to a state in which recovery is realized at the level of symbolic meaning systems through the manifest and latent functions of ritual, thereby producing a shared sense of recovery among people. Even when objective recovery has not yet been achieved, symbolic recovery can serve as a "catalyst" that promotes actual, objective recovery. This is arguably the most compelling aspect of the concept of symbolic recovery (Yama, 2006).

What further sets this perspective apart is its assertion that rituals are not merely spontaneous or culturally inherited phenomena, but are actively "produced" by people to generate a shared sense of symbolic recovery. This theory diverges from traditional anthropological and sociological frameworks of ritual, which tend to emphasize continuity and collective unconsciousness over individual agency.

In this production-oriented view, recovery planners and public officials are not passive actors constrained by the social group but are instead positioned as active agents who intentionally design and implement recovery rituals. These rituals are constructed with the purpose of evoking a sense of recovery within the affected population—not merely as symbolic representation, but as a means to bring about recovery itself (Yama, 2006).

A major obstacle to this theory has been the dominant "ritualism" paradigm in anthropology and sociology, which posits that individuals' actions merely reflect the unconscious will of the group. In this paradigm, even decision-makers such as mayors or recovery planners are seen as mere extensions of their communities. This social-reductionist view assumes that individuals' intentions cannot deviate meaningfully from the collective group consciousness.

However, such an assumption does not hold in all situations. The gap between individual agency and group identity can be significant, even within small communities. Traditional ritualism often ignores this gap, interpreting social phenomena solely through the lens of collective behavior.

To overcome this limitation, the author introduced the concept of "intermediate intellectuals"—individuals who, although embedded within a social group, adopt a semi-external viewpoint. These people are able to access and mobilize external knowledge, skills, and networks, applying them to improve the internal conditions of their communities. By repeatedly interacting with the group, they instigate transformations over time, acting like the dual focal points of an ellipse (Yama, 2020). Such individuals also play central roles in the conception and implementation of recovery rituals.

This reconceptualization of ritual aligns with the ritual theory of Ogyu Sorai, a Confucian scholar from the Edo period. Sorai's theory emphasized the deliberate and creative nature of ritual practices, making it highly compatible with the production-oriented view of symbolic recovery. For a more detailed discussion of Sorai's ritual theory, see Yama (2021).

Based on these theoretical insights, what exactly is symbolic recovery?

In Kobuchihama, a small fishing village in Ishinomaki, Miyagi Prefecture, which was severely damaged by the 2011 Great East Japan Earthquake, only 18 out of 150 houses remained after the tsunami. In July 2016, a summer festival was revived there. Many young people carried *omikoshi* (portable shrines) through the village while shouting festival chants. These young people were not local residents but university students from Tokyo who had volunteered to help revive the festival. The original residents, who had lost their homes to the tsunami, were still living in temporary housing. Hearing the festival chants, elderly residents gathered from the temporary housing areas.

Interestingly, local shrine festivals are typically limited to shrine members, known as *ujiko* (parishioners), and non-members are only allowed to watch, not participate. However, in the wake of the disaster, which resulted in the loss of many homes and lives, outside volunteers were allowed to join. Normally, festivals serve as cultural boundaries to exclude outsiders, but during this crisis, the festival became a "mediating tool" that enabled the inclusion of non-locals.

Similar festival revivals occurred in many coastal communities affected by the Great East Japan Earthquake and tsunami (Takakura & Yamaguchi, eds., 2018).

Beyond festivals, restoration of cultural heritage also plays a central role in symbolic recovery. Another case that illustrates the important role of rebuilding damaged cultural heritage—such as shrines, temples, and traditional structures—in recovery is the 2016 Kumamoto Earthquake.

The Kumamoto Earthquake, which occurred in 2016, resulted in the deaths of 273 people and injuries to around 3,000. A total of 183,882 people were forced to evacuate. Severe damage occurred in towns such as Mashiki, Nishihara, and Minamiaso in Kumamoto Prefecture, where numerous buildings collapsed.

Among the damaged structures was Aso Shrine, whose main gate, designated as an Important Cultural Property by the Japanese government, completely collapsed due to the quake. Kumamoto Castle, another Important Cultural Property, also suffered extensive damage to its roofs, exterior buildings, and stone walls. The castle's *shachihoko* ornaments were destroyed, and many *kawara* tiles fell from the rooftops. Numerous other historical structures were also lost.

These main buildings and structures, particularly religious ones like shrines and temples, serve as important symbols of the local community. As such, many residents felt that community recovery was achieved only when these structures were restored.

Disaster-affected communities often hold collective memories of trauma. The revival of festivals represents a symbolic reenactment of the community as it existed before the disaster and serves as an expression of the spiritual recovery of the local people. These rituals evoke memories of the deceased, symbolically "resurrect" them within the community, and express the ongoing relationship between the living and the dead. Furthermore, such rituals recall the pre-disaster form of the community, acting as if it had returned, and symbolically convey a reconciliation between past and present. It is through the shared wisdom of the community that such collective traumatic memories are healed. In this sense, the idea of symbolic recovery demonstrates the importance and necessity of emotional recovery in the context of disaster recovery.

## 3 Emotional Recovery

Throughout human history, people have enjoyed the blessings of nature while also developing wisdom and strategies to mitigate and cope with disasters brought about by nature, thereby sustaining their lives. The accumulated knowledge, practices, and technologies developed in specific regions to respond to recurring disasters can collectively be referred to as "disaster culture."

Building on this foundation, cultural anthropologist Jiro Tanaka (1986) argued that communities frequently affected by disasters have developed region-specific knowledge and adaptive techniques for interpreting signs of impending danger and for mitigating disaster impacts. These tangible and intangible components—covering the acquisition, transmission, and application of such knowledge—comprise what he terms "disaster culture." Broadly defined, disaster culture includes scientific and engineering expertise, such as earthquake seismology and flood prevention infrastructure, while also encompassing localized perceptions of risk and adaptive behavioral practices (Tanaka, 1986, pp. 9–10).

Isao Hayashi, another anthropologist, identified two principal approaches to studying disaster culture. The first is a descriptive method that documents traditional knowledge and disaster perceptions inherited from historical experiences. The second evaluates the practical relevance of this knowledge for contemporary disaster risk reduction, often referred to as "disaster prevention culture" (Hayashi, 2016, p. 189).

What deserves attention here is the deeper function of disaster culture. Survivors of past disasters often seek to impart their experiences to future generations as lessons, thereby transforming suffering into meaningful narratives and facilitating emotional healing. This practice parallels Claude Lévi-Strauss's insight into traditional societies, where shamans use mythic narratives in healing rituals to symbolically reframe physical suffering, enabling patients to comprehend and accept their pain at a symbolic level (Lévi-Strauss, 1958 [1972]).

For instance, since the story of Noah's Ark depicted in the Old Testament, myths and legends about great floods and humanity's regeneration from survivors have been passed down across cultures around the world. It can be said that humans have sought emotional recovery by transmitting their experiences of disaster to posterity.

In Japan, especially in disaster-stricken areas, the act of passing down experiences as lessons is referred to as *saigai denshō* (disaster storytelling or transmission). These are often conveyed in the form of oral traditions, folktales, proverbs, paintings, and stone monuments, and are sometimes accompanied by ritual events. A representative example is the phrase "*tsunami tendenko*," which gained renewed recognition after the Great East Japan Earthquake. It teaches that when a tsunami is imminent, one must immediately flee to higher ground, without concern even for family members' safety. Other examples include the saying, "When an earthquake strikes, flee to a bamboo grove," or place names that include words like "snake" or "dragon," which are believed to signify areas prone to landslides or erosion.

Another educationally adapted story is *Inamura no Hi* (The Fire of Inamura), which is based on the heroic actions of Goryō Hamaguchi and was later fictionalized. It is also related to Lafcadio Hearn's short story "A Living God."

About 150 years ago, in November 1854 (Ansei 1, according to the lunar calendar), the village of Hiro-mura (present-day Hirogawa Town in Wakayama Prefecture) was struck by the Ansei Nankai Earthquake and a subsequent tsunami. The village suffered extensive damage, with 36 deaths and not a single home spared. Goryō Hamaguchi, then 35 years old, lit fires in stacks of harvested rice sheaves (*inamura*) along the escape route—reportedly from his own family's fields—to help villagers flee to safety. He also worked tirelessly to support recovery efforts by building shelters and distributing tools, and he invested his personal wealth in constructing a 5-meter-high, 600-meter-long embankment (the Hirogawa Embankment), which later protected the village during the 1946 Nankai Earthquake tsunami.

Today, a monument honors Hamaguchi's contributions. Lafcadio Hearn, inspired by both the 1896 Sanriku tsunami and Hamaguchi's deeds, wrote *A Living God* in 1897. This story was later adapted by elementary school teacher Tsunezō Nakai for schoolchildren and published in fifth-grade national textbooks under the title *Inamura no Hi* for about ten years starting in 1937. In this version, an elderly man named Gohei Hamaguchi sees the sea receding, predicts the tsunami, and sets fire to the rice sheaves to alert the unaware villagers and guide them to safety.

Although this adaptation differs from historical fact, it left a deep impression on schoolchildren and was praised for conveying basic disaster preparedness. More recently, during the 2004 Indian Ocean Earthquake and Tsunami, *Inamura no Hi* was revisited as an effective teaching tool for tsunami awareness (MLIT White Paper, 2004).

Monuments that transmit disaster experiences, the remains of destroyed buildings, and memorials for victims—known as "disaster heritage" or "disaster ruins"—also draw attention. For example, in the Kurokami area of

FIGURE 1.1 Kurokami Buried Torii Gate, Sakurajima
SOURCE: PHOTO BY THE AUTHOR

Sakurajima, Kagoshima City, there is the "Kurokami Buried Torii" of Haragosha Shrine (Figure 1.1). This torii gate was almost entirely buried by pumice and volcanic ash during the 1914 Taisho eruption, with only its top now visible above ground. It is said that, by the wise decision of the village head at the time, the scene was preserved as it was immediately after the eruption, to serve as a memory for future generations. More than 110 years have passed since the Taisho eruption, and with the possibility of another large-scale eruption in the near future, the value of such a disaster heritage site—which vividly conveys the challenges posed by heavy volcanic ashfall—is considered extremely high. There is a growing call for it to be effectively utilized for disaster preparedness (Yama, 2022).

In the Tohoku region, which has repeatedly suffered massive tsunami disasters in the past, many monuments and shrines remain to preserve the memories of such disasters. One example is the "Great Tsunami Memorial Stone" in Miyako City, Iwate Prefecture, which records the damage from the tsunami caused by the 1933 Showa Sanriku Earthquake and bears the warning, "Do not build houses below this point."

Philosopher Ching Yuen Cheung reflected on an important realization during his visit to Namiwake Shrine in Sendai City, Miyagi Prefecture. He noted that this shrine embodies the memory of the earthquake and tsunami, and that visiting such a place—undertaking a kind of "pilgrimage"—serves to revive that memory (Cheung, 2022). Inspired by the work of Japan's leading philosopher Tetsurō Watsuji (1889–1960), particularly his book *Pilgrimage to the Ancient Temples*, Cheung has engaged in what he calls "pilgrimage to disaster sites" in the Tohoku region affected by the Great East Japan Earthquake. Visiting disaster heritage and ruins is a way to retrieve and pass down such memories.

Today, this concept of pilgrimage to sites of disaster memory has become institutionalized in the form of museum-style programs. For example, the "Kumamoto Earthquake Corridor of Memory" is a corridor-style field museum that networks preserved disaster heritage and ruins across the affected areas of the 2016 Kumamoto Earthquake. Visitors can learn from these sites and participate in the transmission of disaster memory.

Disaster remains from earthquakes are now often referred to specifically as *shinsai ikō* (earthquake disaster remains)—a term that came into common use after the Great East Japan Earthquake. In the affected areas of that disaster, many such remains have been preserved, and museums have been established to ensure that these memories are not lost.

The Geospatial Information Authority of Japan has also created a new map symbol for *natural disaster memorial markers* (*shizen saigai denshōhi*) and begun marking them on maps. Japan has experienced numerous natural disasters throughout history. Each time, people inscribed the lessons learned into stone monuments and left them for future generations. However, many people today are unaware of the existence of these monuments, or even if they are, they may not know what is written on them. As a result, the lessons of the past are not passed down, and the same kinds of disasters recur. By mapping these markers, the Geospatial Information Authority hopes to preserve historical lessons and reduce future damage.

In many local districts, residents have built memorial monuments, often accompanied by annual memorial ceremonies or festivals. These communities carry shared psychological trauma. The act of constructing memorials and holding such events can be seen as a symbolic gesture toward the recovery of the community's emotional well-being. It provides a path for communities to move forward.

In Kobe, one of the areas affected by the 1995 Great Hanshin-Awaji Earthquake, an annual event called "Kobe Luminarie" has been held every winter since the disaster. Its purpose is to mourn the victims and ensure that the memory of the earthquake is passed on to future generations.

In the southern Kanto region, which suffered extensive damage in the 1923 Great Kanto Earthquake, many memorial monuments and towers have been built to honor the victims. These structures often serve religious functions and are accompanied by memorial services. One of the most well-known is the Tokyo Metropolitan Memorial Hall located in Yokoamicho Park. Built on the site of the former military clothing depot—the area most severely affected by the earthquake—it houses the remains of the victims. A memorial ceremony is held there every year on September 1, the anniversary of the earthquake. Within the same park stands the Tokyo Reconstruction Memorial Hall, which functions as a museum to preserve the memory of the disaster.

On September 1, 2023, the 100th anniversary of the Great Kanto Earthquake, the author had the opportunity to visit the Tokyo Metropolitan Memorial Hall. Many people came to offer prayers. Nearby, in front of the memorial for Korean victims of the earthquake massacre, a commemorative ceremony was held. Particularly striking was a memorial dance performance of Korean traditional dance. The following day, on September 2, 2023, the author visited the "Memorial to Korean Victims of the Great Kanto Earthquake" located under the Arakawa embankment in Yahiro, Sumida Ward, Tokyo. There, a memorial service was held for Koreans who were massacred by Japanese civilians acting on false rumors. Once again, a performance of samulnori, a traditional Korean performing art, left a deep impression (Figure 1.2).

The deaths resulting from the massacre are fundamentally different from those caused directly by the earthquake or subsequent fires. While the latter were due to a natural disaster, the former were the result of a human-perpetrated crime. It is inappropriate to treat the victims of these two causes as equivalent. Subsuming both groups under a single category of "victims of the Great Kanto Earthquake" risks concealing the existence of the massacre and its victims—as evidenced by the attitudes of the Japanese government and Tokyo Metropolitan Government.

What is truly needed is the emotional recovery of the victims of the massacre, their families, and related communities. Conversely, unless their emotional healing is achieved, true recovery from the Great Kanto Earthquake will remain incomplete. For that to happen, society must acknowledge the victims of the massacre and ensure that the memory of the incident is preserved.

This is why deeply moving rituals—especially those that incorporate traditional performing arts or ethnic music—are essential. The author was particularly touched by a reading session jointly organized by young Zainichi Koreans and Japanese youth. Through readings of testimonies and personal reflections on the massacre, the participants gave voice to emotions that resonated strongly with attendees. Such acts can be seen as contemporary forms of memorial and commemorative rituals. These ritual practices not only play a central role in preserving memory but are also crucial for emotional recovery.

FIGURE 1.2 Samulnori performance during the memorial ceremony for Korean victims of the Great Kanto Earthquake at the Arakawa riverbank
SOURCE: PHOTO BY HONGKI LEE

## 4 The Disaster-Affected Hometown and the Restoration of *Fudo* (Natural and Cultural Climate)

On January 1, 2024, the Noto Peninsula Earthquake struck Japan. However, the affected areas are still facing difficult living conditions, and many people have been forced to continue their lives in evacuation shelters. One of the notable characteristics of the Noto Peninsula Earthquake is that it was a large-scale disaster that struck a depopulated rural area. Not only the Noto Peninsula, but many areas frequently hit by major disasters have long suffered from population decline and aging—what is often referred to as *kaso*, or depopulation. If we consider *kaso* as "another kind of disaster" that gradually weakens communities, then it can be said that many depopulated areas have already been in a state of ongoing disaster even before any natural catastrophe occurred (Okada, 2015). When such areas are hit by large-scale disasters, they face a compounded or multilayered disaster situation. The difficulty in recovering from disasters in depopulated areas lies in this condition of layered affliction.

In such regions, it is necessary for local residents to proactively recognize the vulnerabilities of their communities, prepare for future disasters, and initiate efforts toward post-disaster recovery in advance.

In many depopulated areas, the outflow of population to urban centers has been ongoing. For the children and grandchildren who live in cities, these regions are their ancestral hometowns where their elderly parents or grandparents still reside. Whether one continues to live in the region or has left it, such places hold deep emotional meaning. When a depopulated area is struck by disaster, it means that a hometown—a source of emotional support for many—is also damaged. Therefore, the psychological shock of a hometown being devastated is profound, not only for those currently living there but also for those who have moved away.

In the recent earthquake, many cases were observed where children and grandchildren, who had returned home for the New Year's holiday to visit their elderly parents, experienced the disaster together with their families in their hometown. This situation was unique precisely because it occurred in a place considered a *hometown*. Furthermore, when a hometown is affected by disaster, it also means that elderly parents and grandparents who remained in the region are impacted. Preexisting concerns over their well-being become a harsh reality, prompting some to relocate their family members to the city. As a result, some villages may experience accelerated depopulation in the aftermath. Thus, when a depopulated region is affected by disaster, it is not just about a place with declining population and aging residents being hit—it must also be understood through the lens of *hometown disaster*. Without this perspective, the challenges of recovery in such regions cannot be fully grasped.

Depopulated areas are also defined by their unique *fudo*—a rich integration of natural environment, historical lifestyle, livelihoods, and culture cultivated over generations. The disaster, therefore, impacts not only people's lives but also this composite cultural and environmental identity. Recovery in such areas must mean not only the reconstruction of the hometown but also the restoration of its *fudo*.

According to the philosopher Tetsurō Watsuji (1889–1960), in his work *Fudo: A Study of the Human and Climatic Environment* (1935), *fudo* is not merely the natural environment but rather an "expression of human existence as a subjective being." In that sense, the "reconstruction of *fudo*" is also an expression of the recovery of the people who live there—in essence, the "recovery of the human."

In fact, the restoration of *fudo* symbolically represents the recovery of the identity and spirit of the affected community. The Noto Peninsula is home to a wealth of such natural and cultural heritage, including the scenic Senmaida rice terraces of Shiroyone, designated as part of the Globally Important Agricultural Heritage Systems (*Noto's Satoyama and Satoumi*), the traditional streetscapes

FIGURE 1.3 The Wajima Morning Market. One of the oldest traditional markets in Ishikawa Prefecture, photographed after the 2024 Noto Peninsula Earthquake
SOURCE: PHOTO BY THE AUTHOR

of the Wajima Kuroshima preservation district, the ritual "Aenokoto" of Oku-Noto recognized as UNESCO Intangible Cultural Heritage, and the "Noto Kiriko Festival" with its massive lantern processions.

The Wajima Morning Market, with over a thousand years of history, was lost in the large fires following the earthquake (Figure 1.3). Reviving the globally significant *fudo* of the Noto Peninsula—where nature, culture, and history are deeply intertwined—will undoubtedly play a major role in the recovery of the disaster-affected region.

## 5 Emotional Recovery and the Philosophy Café

One type of activity that has become increasingly common in disaster-affected areas of Japan is the philosophy café. In Paris, France, people gather at cafés on Sunday mornings to engage in open conversations on various topics in what is known as the "Café Philosophique." Anyone can participate simply by paying for a drink—no introductions are required. Those who wish to speak can express any opinion, while others may choose only to listen. There is no need to reach conclusions or consensus; the goal is simply for participants to share their diverse views. The first philosophy café is said to have been launched by the philosopher Marc Sautet on December 13, 1992, at the Café des Phares near the Place de la Bastille.

In Japan, the philosopher Kiyokazu Washida is considered the pioneer of philosophical dialogue in response to disaster. After experiencing the Great Hanshin-Awaji Earthquake on January 17, 1995, Washida began engaging in philosophical conversations with the belief that philosophers had a duty to stand alongside disaster survivors. The momentum for philosophy café activities gained renewed traction after the Great East Japan Earthquake on March 11, 2011. In Fukushima, which was affected by the nuclear accident, as well as across tsunami-affected areas in the Tohoku region, philosophy cafés became increasingly active. One major hub for such events was Sendai Mediatheque, located in the disaster-hit city of Sendai (Washida, 2014).

Furthermore, the spread of philosophy cafés throughout Japan accelerated with the global outbreak of COVID-19. With the rise of online meetings, philosophy cafés began to be held regularly at hundreds of locations across the country.

From April 2013, the author spent a year studying abroad in France and encountered philosophy cafés in Paris. Instinctively recognizing their potential for community development, the author introduced philosophy cafés in various places after returning to Japan—particularly in regions affected by the Hanshin-Awaji Earthquake (Figure 1.4). This initiative was driven by the author's long-standing engagement in practical, collaborative research on community development and post-disaster/regional revitalization, involving residents, governments, and nonprofit organizations. Through this work, it became clear that many communities lacked spaces where people could gather and speak freely about various topics.

In many local communities, it is difficult to discuss "serious" topics—such as social issues, local problems, or even personal concerns—because both speakers and listeners are unsure how to respond, making such discussions socially awkward or taboo. As a result, many people have no experience talking about serious matters, or any opportunity to learn how to engage in such dialogue. Without a shared experience of discussion, it becomes difficult to develop a shared sense of concern or take collective action toward problem solving. To address this issue, the author has introduced philosophy cafés across Japan (Yama, 2022).

At philosophy cafés, participants listen intently with serious expressions. Sometimes laughter or murmurs arise, and in a warm and friendly atmosphere,

FIGURE 1.4 The Philosophy Café
SOURCE: PHOTO BY THE AUTHOR

a round of applause often follows when someone finishes speaking. Themes vary from session to session, beginning with "knowledge" in the first gathering, followed by topics such as "work," "play," and "common sense." For example, the 17th session focused on "disaster." This theme emerged naturally, spontaneously, and the session became a valuable opportunity for participants to speak freely and share their concerns.

Through the philosophy café, many participants have acquired the skills of dialogue and gone on to initiate diverse forms of community activity. Creating such spaces for open conversation is now considered a crucial aspect of community development. For researchers as well, philosophy cafés offer a valuable opportunity to hear the voices of local residents. These gatherings allow researchers to understand people's emotions and learn their "language." At the same time, they are sites where residents and researchers learn each other's ways of expression and collaboratively create new forms of language.

The development of philosophy cafés in Japan began with the Hanshin-Awaji Earthquake, spread widely in the aftermath of the Great East Japan Earthquake—especially in disaster-affected areas like Fukushima—and expanded nationwide, particularly through online platforms during the COVID-19 pandemic. As this trajectory suggests, philosophy cafés in Japan should be understood as an initiative seeking to reflect the voices of disaster survivors in the process of recovery.

## 6 Conclusion

The importance of dialogue, as discussed above, was reaffirmed in the wake of the Noto Peninsula Earthquake, where numerous isolated settlements emerged and took a long time to be reconnected. One of the underlying reasons for this isolation is that many depopulated villages in rural areas were already, even during ordinary times, in a state close to isolation. If this is the case, then it is reasonable to expect that such communities will become isolated during disasters—and preparations must be made with this in mind.

In addition to securing evacuation routes, food and drinking water, water for daily use, alternative communication methods besides telephones, backup power sources during outages, and support networks from outside, it is essential for local residents to engage in proactive discussions beforehand to ensure mutual cooperation after a disaster. It is also important for residents to talk about what is necessary to continue living in the area, what can be done to sustain the community, and what they envision for its future. Without such conversations, communities cannot share their challenges or cooperate in taking collective action toward solving them.

However, many depopulated areas are already on a trajectory toward disappearance without having taken any such preparatory steps. Should a disaster strike under these conditions, depopulation may accelerate rapidly, and some settlements may, in fact, vanish entirely. If *kaso* (depopulation) is seen as "another form of disaster," then it can be said that many areas are already experiencing "hometown disasters" due to depopulation—and are at risk of moving swiftly toward the "loss of hometowns."

To prepare for future disasters, engaging in dialogue before a disaster so that people can cooperate and support each other afterward—and also to discuss the future of the community itself—will likely serve as a critical turning point for the long-term viability of depopulated rural settlements.

## Bibliography

Anderson, B. (1991). *Imagined Communities: Reflections on the Origin and Spread of Nationalism*. London: Verso. (= 1997, 白石さや他訳『増補 想像の共同体——ナショナリズムの起源と流行』東京：NTT 出版.)

Cheung, C.Y. (2022). 物語與日本哲學―哲學的民俗學轉向. Taipei: Wunan Books Publishing.

Cohen, A.P. (1985). *The Symbolic Construction of Community.* New York: Tavistock Publications. (= 2005, 吉瀬雄一訳『コミュニティは創られる』東京：八千代出版.)

Durkheim, É. (1912). *Les formes élémentaires de la vie religieuse: le système totémique en Australie.* Paris: F. Alcan. (= 1975, 古野清人訳『宗教生活の原初形態』東京：岩波書店.)

Fukuda, T. (2012). 復刻版 復興経済の原理及若干問題 (Edited by S. Yamanaka & T. Inoue). Nishinomiya: Kwansei Gakuin University Press.

Geertz, C. (1980). *Negara: The Theatre State in Nineteenth-Century Bali.* New Jersey: Princeton University Press. (= 1990, 小泉潤二訳『ヌガラ：19世紀バリの劇場国家』東京：みすず書房.)

Hayashi, I. (2016). "Saigai Bunka" [Disaster Culture]. In Murosaki, M., Okada, N., Nakabayashi, K., Noro, M., Tsukui, S., & Yamazaki, E. (Eds.), *Saigai Taiō Hando Bukku* [Disaster Response Handbook] (pp. 188–191). Tokyo: Hōritsu Bunka Corporation.

Hyogo Prefecture. (2005). 阪神・淡路大震災の死者にかかる調査について [Survey on the deceased from the Great Hanshin-Awaji Earthquake].

Kumamoto Prefecture. (2018). 震災関連死の概況について [Overview of disaster-related deaths].

Lévi-Strauss, C. (1958). *Anthropologie Structurale.* Paris: Plon. (= 1972, 荒川幾男他訳『構造人類学』東京：みすず書房.)

Miyahara, K. (2006).「『復興』とは何か―再生型災害復興と成熟社会」[What is "Fukkou"? Regenerative disaster recovery and a mature society]. *Sentan Shakai Kenkyu*, 5.

Nagamatsu, S. (2020).「復興とは何か：日本災害復興学会「復興とは何かを考える委員会(2009-2011)」の経緯と成果」[What is Recovery? Summary of the committee of the Japan Society for Disaster Recovery and Revitalization]. *Japan Society for Disaster Recovery and Revitalization Journal*, 15.

Okada, N. (2015). ひとりから始める事起こしのすすめ [Starting Change Alone: A Zero-Based Challenge in Regional Revitalization]. Nishinomiya: Kwansei Gakuin University Press.

Reconstruction Agency. (2012). 東日本大震災における震災関連死に関する報告 [Report on disaster-related deaths in the Great East Japan Earthquake].

Shiozaki, K. (2014). 復興〈災害〉―阪神・淡路大震災と東日本大震災 [Reconstruction as Disaster: The Hanshin-Awaji and Great East Japan Earthquakes]. Tokyo: Iwanami Shoten.

Takakura, H., & Yamaguchi, M. (Eds.). (2018). 震災後の地域文化と被災者の民俗誌 [Local Culture and Ethnography of Survivors after the Disaster]. Tokyo: Shinsensha.

Tanaka, J. (1986). "Saigai to Ningen" [Disasters and Humans]. In Tanaka, J., Tanaka, S., & Hayashi, H. (Eds.), *Saigai to Ningen Kōdō* [Disaster and Human Behavior] (p. 235). Tokyo: Tōkai University Press.

Turner, V.W. (1969). *The Ritual Process: Structure and Anti-Structure*. London: Routledge & K. Paul. (= 1996, 冨倉光雄訳『儀礼の過程』東京：新思索社.)

Washida, K. (Supervisor), & Café Philo (Eds.). (2014). 哲学カフェのつくりかた [How to Create a Philosophy Café]. Osaka: Osaka University Press.

Watsuji, T. (1919). 古寺巡礼 [Pilgrimage to Ancient Temples]. Tokyo: Iwanami Shoten.

Watsuji, T. (1935). 風土ー人間学的考察 [Fūdo: A Study of the Human and Climatic Environment]. Tokyo: Iwanami Shoten.

Yama, Y. (2006).「『象徴的復興』とは何か」[What is Symbolic Recovery?]. *Sentan Shakai Kenkyu*, 5.

Yama, Y. (2020).「『媒介的知識人』とは何か」[What is an "Intermediate Intellectual"?]. *Disaster Recovery Studies*, 11.

Yama, Y. (2021). Sociology of Ritual and Narrative as Post-Western Sociology: From the Perspective of Confucianism and Nativism in the Edo Period of Japan. *The Journal of Chinese Sociology*, 8, Article 8.

Yama, Y. (2022).「語り合う場のデザインー哲学カフェの試みから」[Designing Dialogical Spaces: The Philosophy Café Approach]. *DPRI Newsletter*, 101, 7. Kyoto: Kyoto University Disaster Prevention Research Institute.

Yama, Y. (2022).「災害遺構の活用を考えるー桜島の大正大噴火の黒神埋没鳥居を訪ねる」[Utilizing Disaster Heritage: Visiting the Buried Torii of Sakurajima's 1914 Eruption]. *FUKKOU*, 49.

Yama, Y. (2023).「関東地震の災害遺構—慰霊・追悼と記憶の継承」[Disaster Heritage of the Great Kanto Earthquake: Memorials, Mourning, and the Transmission of Memory]. *FUKKOU*, 52, 6. Nishinomiya: Kwansei Gakuin University Institute for Disaster Recovery System Research.

Yamanaka, S. (2023). 人間の復興 [Human Recovery]. Nishinomiya: Kwansei Gakuin University Press.

CHAPTER 2

# Sociology and Disasters in Europe and in East Asia

*Laurence Roulleau-Berger*

The global ecological crisis is increasingly manifesting itself in different regions of the world, here in East Asia and Europe, at unpredictable intervals in the form of unprecedented natural, health, social, political, and economic disasters that vary according to societal, cultural, and civilizational contexts (Li, Roulleau-Berger, 2016). These destructive phenomena produce localized and multi-localized inequalities on a global scale, drawing new geographical, climatic, social, and moral boundaries within the framework of an "ecological cosmopolitanism." Individuals also develop capabilities for survival, recovery, and mobilization in the post-disaster process. Democratic and authoritarian, democratic-authoritarian, or authoritarian-democratic governments put in place biopolitical mechanisms for societal reconstruction that will favor, to varying degrees, the production of subpolitical narratives by disaster victims. Democratic regimes will favor forms of egalitarian cooperation between political and institutional actors and citizens, while authoritarian regimes will impose greater collective discipline to reorganize social life. In this chapter, we will present the main trends in sociological research on disasters in China, Japan, and Western Europe. We will first see that in Western Europe, environmental sociology has been built around a Western way of thinking about the relationship between risk, disaster, and modernity. We will then select important works that show how Japanese environmental sociology has produced highly situated knowledge and autonomous knowledge, and how research by Chinese sociologists and anthropologists has also highlighted non-Western approaches to disasters. Then, from a post-Western sociological perspective, we will present various works—a selection that is in no way exhaustive—based on theoretical axes that have mobilized researchers in East Asia and Europe:

- disasters, memory, and emotions
- disasters' biopolitics, interactional citizenship, and trust
- capabilities and mobilizations

## 1 Western Environmental Sociology, Risks, and Modernity

Environmental sociology, which includes disaster studies, began in the 1980s and 1990s with the emergence of environmental problems. In Europe,

 HTTPS://DOI.ORG/10.1163/9789004751804-004

the reflexive modernization theories of Anthony Giddens and Ulrich Beck inspired the notion of *ecological modernization* (Mol, Spaargaren, 2000; Mol, Sonnenfeld, 2000) as "the social scientific interpretation of environmental processes at multiple scales in the contemporary world." According to Guillaume Faburel (2016), over the last forty years we have been confronted with ecological irreversibilities in capitalist societies as well as a de-synchronization between social activities, places of decision-making, and collective action; on one hand, a renewed conception of the environment is more and more "fractal" and active, but on the other hand, environments involve citizens in new experiences of unequal situations.

In "We Have Never Been Modern," Bruno Latour proposed abandoning the old distinctions inherent in Western "modern" thinking—particularly between nature and society—in order to propose a different sociology. This perspective led him to give a central place to ecological crisis in his thinking. Bruno Latour therefore invited us very early on to free ourselves from Western thinking in order to think about catastrophes and disasters. According to Ulrich Beck and Edgar Grande (2010) European societies have entered a second phase of modernity, a reflexive modernity, characterized by the need to simultaneously face processes of individualization, underemployment, and financial and environmental risks in the global society, but Ulrick Beck is speaking about Western societies without taking into account the variety of non-Western modernities. Shmuel Eisenstadt (2000) produced the theory of "multiple modernities" to describe through a comparative civilizational perspective the plural forms of modernities in diverse historical and structural contexts. Göran Therborn (2003) has considered the different varieties of modernity and non-modernity; he looked at the ways these do not simply coexist and challenge each other, but also at how they are *entangled* with each other in various ways. But Shamuel Eisenstadt and Göran Therborn did not introduce the question of ecological risks in the theories of non-Western modernities.

Western sociology of catastrophes and disasters did not really engage in dialogue with the sociology of modernity, even when it became internationalized in the 1970s. Its center of gravity remained in the United States with the creation in 1984 of the Research Committee on Disasters within the International Sociological Association and its journal, the *International Journal of Mass Emergencies and Disasters*. It was not until the 1990s that the sociology of disasters took shape in France (Giry, 2023). According to Benoît Giry, in Western sociology, disasters are mainly thought of in terms of the collective suffering that constitutes an "event" and the establishment or even maintenance of a pathological social order, rather than in terms of a break with what preceded them. The promoters of *critical studies* considered that disasters could appear as an "interpretative fiction," suggesting that they could appear to disrupt the "normal course of things" (Giry, 2023).

In the 1970s and 1980s, research in the social sciences and humanities was devoted primarily to major technological and natural risks, the hidden side of risk, it means *the fabric of the risk*, it also means what happens in "discrete spaces" and what appears spectacular in public and political space (Gilbert and Henry, 2012). Torny and Chateauraynaud (1999) promoted the idea that certain actors can become *lanceur d'alerte* and reveal the invisibility of a problem. Sandrine Revet (2013) proposed putting an end to the "culture of risk" which, having enjoyed growing success in the 1990s, became a "culture of prevention" or a "subculture of disaster" to think simultaneously, and even in tension, about biopolitical government mechanisms on the one hand, and ordinary, networked practices and daily routines of disasters on the other. Facing new ecological risks or disasters reveals the maintenance of previous inequalities, the production of new ones, and the breaking away from previous ones. It also means dealing with the right to the appropriation and production of space on ecological scales and in historical places; Guillaume Faburel has shown how the metropolization of the world through a dual movement of diversification and expansion of cities is producing an *ecocide* with the depletion of resources and the massive destruction of all ecosystems (Faburel, 2020).

Moreover, the concept of *environmental justice*, which emerged in the United States in the 1980s, when African American, Latino, and Native American communities protested against *environmental racism* (Bullard, 1993), has become increasingly central (Nygren, 2014). It refers to inequalities in the spatial distribution of risks; some authors propose that we consider the interaction between environmental justice and vulnerability (Walker, 2009). However, according to Cary and Villalba (2023), there is still little research on environmental racism in France, due to its colonial past.

Since the 1990s, in a context of ecological transformation, following various food crises in Western countries, food studies have been developed in sociology of environment and around the theory of strategic warning signals. The sociology of food studies shows food anxiety is an anthropological invariant (Fischler, 2003; Poulain, 2013). Jean-Pierre Poulain introduced food security, fraud crisis, food safety, and controversial risk to think democratic risk management to the social reconstruction of food in social arena crisis. More recently, in the face of unequal access to food, the concepts of *food justice* and *food sovereignty* have been introduced in food studies; Camille Hochedez and Julie Le Gall (2016) proposed to theorize, foster, and promote an *agri-food justice* which should improve the access to better food within the whole food systems.

In the early 1990s, heated scientific controversies emerged around the Anthropocene, which appeared as a Western-centric narrative producing a hegemonic vision of the history of different peoples in their relationship with nature, a narrative constructed around binaries, such as the nature/culture dichotomy. For some authors, the Anthropocene discourse is another version of the capitalist white male who finds in this narrative a renewed way to impose his neo-colonial domination; for Jason Moore, the term Anthropocene should be replaced with the concept of Capitalocene, or even *racial Capitalocene* (Vergès, 2017), which means the imposition of domination, colonialism, and racism that produce ecological disasters; what is being questioned here are the Western life-styles of our consumerist and overproducing societies (Cary, Villalba, 2023).

Finally, if the theory of Anthropocene has taken its central status in Western environmental sociology, for Paul Jobin (2023) an overwhelming proportion of authors who discuss the Anthropocene, both advocates and opponents of the notion, are from Western Europe, North America, and Australia. Departing from the domination of Western paradigms in the Anthropocene literature, a few Asian scholars have therefore deemed it necessary to redefine the concept from the angle of postcolonial criticism.

## 2 Environmental Sociology and Autonomous Knowledge in East Asia

While a sociology of disasters has been developed within the framework of hegemonic Western sociologies, and concepts such as vulnerability and resilience have been imported from Western sociologies into non-Western sociologies, their use has not promoted the visibility of situated theories that account for disasters in regions of the world, such as East Asia or Southeast Asia. While Kathleen J. Tierney (2019) proposed distinguishing between emergencies, disasters, and catastrophes, sociologists in East Asia tend to refer to *disaster studies*. This usage refers to the scale and intensity of disasters in China, Japan, and Indonesia, which have not been experienced as such in Europe. In Chinese and Japanese sociology, social, cultural, political, and moral continuities and discontinuities are thought of in terms of a permanent dialectic between a "before" and an "after" the disasters; socialization, desocialization, and resocialization are thought of as simultaneous, whereas they are rather disjointed in Western sociology.

### 2.1 *The Unique Perspectives and Contributions of Japanese Environmental Sociology*

In Japan, Keiichi Satoh and Hidehiro Yamamoto (2023) considers ecological modernization theory (EMT) as one of the major environmental theories/

discourses among scholars and policymakers. The EMT draws its perspective mainly from the Western countries that experienced the set of the characteristics of late modernity, such as post-industrialization and severe environmental degradation. By contrast, the East Asian countries generally experience the modernization and the second modernization that result in East Asia's unique response to the environmental crisis. East Asian response to environmental risk is based on the modern, authoritarian, and technological perspective and not from the reflexive modernity.

According to Koichi Hasegawa (2023) we can consider the unique perspectives and contributions of Japanese and non-Western environmental sociology. For him the influences of Western sociology on environmental Japanese sociology are not visible. The Japanese sociologists have developed original perspectives, including the social structure of victims (Iijima, 1993), life environmentalism (Torigoe and Kada, 1984), industrial pollution incidents, intersections of daily life and culture with ecology, disasters affects, and recovery processes. Excellent contributions have been realized by Iijima, Torigoe, and Funabashi and al. in 1985, who have produced benefit versus victimized zone theory. Koichi Hasegawa (2004) has developed a *sociology of the downstream perspective* on daily life production, residents' movements, and social structure of victims; here, the downstream approach is focused on a conceptualization of society in terms of sustainability and resource circulation. In a closed perspective, Yoshiyuki Yama (2006), after exploring the concept of *fukkou* (to be revived again), which provoked disputes in the academic world, has introduced the concept of *symbolic recovery* from an anthropological perspective, which means a focus on the emotional healing, restoration, and revitalization of people.

Since the Great East Japan earthquake in 2011 represents a turning point in Japanese sociology, disaster sociology has been well developed and is focused on a fundamental perspective in sociology, the interrelationship between the individual and society (Satoh, Yamamoto, 2023). For Shigeyoshi Tanaka (2023) nature can no longer be regarded as existing outside of society, natural disasters do not exist in a natural world, and disasters should be defined as continuous processes occurring from the pre-disaster period through the reconstruction period. So, Shigeyoshi Tanaka considers "disaster" in the context of both a social structure and a social change to escape the marginal status of disaster sociology in sociology; for him a disaster can be seen as a "snapshot" of a society, and can be a means to produce knowledge on the relationship between society and natural disasters in the future of sociology.

Japanese scholars (Chen, Sugiman, 2010; Nagata, 2012, 2016) have been focused on what constitutes the "re-creation of society" after a huge disaster, what creates resilience, and how citizens learn something new and adapt themselves to a similar disaster. Adaptive learning means resocialization, the production of new inequalities, and new identities. Then, the question of the social construction of anxiety has been introduced. For example, in food studies Niyama (2012) wondered what scientific communication produced by experts and government is provided to citizens to ease anxiety; Yamaguchi (2019) analyzed the self-protective behavior of egg producers in Japan when they faced risks generated by avian influenza, and the compensation provided by the government.

In Japanese and Korean sociologies we can select some common knowledge. For example, Hong Deok-hwa and Ku Dowan (2023) are revealing how Korean environmental sociology can contribute to broadening the understanding of the relationship between ecological transition, environmental movement, state, and democracy. Through the development path of Korea's environmental movement, it is also possible to trace another historical path of environmentalism in social conditions different from those of Western environmentalism or environmentalism of the poor.

Japanese sociologists do not view disasters as turning points or points of rupture in societies, but rather as a continuum between pre-disaster, disaster, and post-disaster, i.e., how communities of victims survive, reform, and rebuild. Rina Kojima (2023) has developed a theoretical framework built around the articulation between structural violence in the reconstruction following a disaster and the symbolic recognition of the victims. Shujiro Yazawa (2014) developed a theory of risk culture rather than risk society after the Great East Japan Earthquake and Tsunami Disaster, to understand non-institutional, emergent sociation, horizontal orders, and reflexive communities.

### 2.2 *Some Tendencies in Chinese Environmental Sociology and Anthropology*

In China it was after the mid-1990s that the sociology of the environment was introduced as a discipline. According to the "China Sociology Yearbook", between 2011 and 2015 there were 519 academic research works on the themes "environmental sociology" and "environment and society."

Hong Dayong (2011, 2017, 2023) has analyzed the social factors of environmental destruction in contemporary China:

- the transformation of the social structure by industrialization, urbanization, and regional differentiation;
- the transition of economic and political systems, and the market economy;
- decentralization, reform, and change of control;
- news values with moral decline, consumerism, and the intensification of social mobility.

According to Zhang Yulin (2019), the last 40 years of urbanization have led to a massive consumption of ecological resources, causing disasters and urban pathologies; a body of research has shown how rapid urbanization, industrialization, and economic liberalization—that is, *compressed modernity* (Chang, 2010)—has led to the emergence of environmental problems (Bao, Chen, 2011). Bao Zhiming and Zeng Wenqiang (2021), based on a study of a village called Y., showed how overexploitation of natural resources had caused serious ecological problems, which were managed in different stages by the local government with the participation of villagers to produce green ecological agriculture (Bao, Yan, 2022).

Several works highlight the relationships between political regime, institutional failures, complicity between local governments, companies, and environmental degradation (Zhang, 2006).

In this perspective Chen Ajiang (2015) demonstrated that the imbalance of the power of the stakeholders, the bureaucratization of the rural grassroots organizations, and the disappearance of the self-organization of the villagers, as well as the loss of traditional ethical norms in the community, are the main causes of the water pollution. Zhang Yulin (2021) employed the theory of the Anthropocene to address the extinction of species and the ecological crisis of the Yangtze River, which was caused by a series of human activities in the context of compressed modernity. According to Zhang Qiaoyun (Zhang, Barrios, 2017) Chinese disaster anthropology did not select the Western theory of Capitalocene and environmental racism to emphasize "cultural reconstruction," which means the "reconstruction of everyday life."

Although we should not confuse risk society and catastrophe society, we can observe some quite similar processes affecting *social* vulnerability, inequalities, and individual and collective capabilities. In Asia social inequalities and natural inequalities are fusing into compressed modernities linked to spatial and historical processes of urban ecological change. Natural catastrophes especially reflect the social vulnerability of certain countries and social groups; disaster is to be understood as a specific moment within social and societal processes and will be viewed as a socio-historical process of production and distribution of social fragility.

Concerning East Asia, Ulrich Beck (2013) spoke of a cosmopolitan risk community (or "Cosmo-Climate"). He distinguished "cosmopolitan empathy" and the subpolitics of "cosmopolitanism from below". In China Li Peilin, in Japan Shujiro Yazawa, and in Korea Chang Kyung-Sup consider that we cannot conceive the risk issue in the same terms in Europe and in Asia because different compressed modernities mean that the complexities, diversities, and heterogeneities of risks that people in East Asia face today overtake those faced by the Western world.

## 3 Disasters, Memory, and Emotions

Japanese anthropologists have produced situated knowledge in developing different approaches of disaster and memory. For Yoshiyuki Yama (2023) disaster-affected communities often hold memories of trauma, so the restoration of local symbols as a vital part of the recovery process looks important. *Fudo* (natural and cultural climate)—a notion produced by Tetsuro Watsuji—means an expression of human existence as a subjective being. For Nomiya Daishiro (2023), away from the Western ethos, a "*hibakusha*" collective memory represented a significant turning point in the postwar history of two cities affected by the atomic bomb, Hiroshima and Nagasaki; it makes a counterargument against Western modernity and capitalism.

In China, Luo Hongguang and Liu Zhengai (2016) simultaneously took into consideration structural inequalities and community reconstruction, the re-creation of society, and the "emergency rescue" and "recovery." They articulate social structures in pre- and post-disaster situations with cultural creativity; they have studied in a post-disaster context the continuity of traditional social support networks after a disaster as well as new social relations, the issue of subjectivity, and the role of folk religious practices (Liu, 2009), the changes of meaning in labor spaces and life spaces (Zhao, 2007). Jin Chen (2016), in a closed perspective, analyzed how cultural norms, religious beliefs, and indigenous knowledge could inform the production of rituals during and after disasters; the mobilization of the concept of indigenous knowledge allows us to understand the complexity and the diversity of the victims and to hear their voices. And the relationship between disaster and rituals is further contingent upon broader socio-economic and political contexts.

Laurence Roulleau-Berger (2016) considered communities of destiny (Pollak, 1998; Barbot, Dodier, 2010), which means emotional communities which are formed in disaster situations and produce ways of being together which will continue during the process of reconstructing the towns and villages. In different geographical and political contexts, these emotional communities form upon the basis of the ordeals of life and shared norms for the reconstruction of a shared world. Facing bereavement and visiting tombs are strong components in the definition of these emotional communities. They are formed from social links upon which the political and religious context and the local cultures have a partial effect.

Paula Vasquez (1999), in her academic work, included the political anthropology of the state in Latin America, the anthropology of disasters and extreme situations, and the anthropology of political emotions. In her work she privileges the individual experience and the subjective construction of the people's relationship with the state and the extreme.

Facing risks also implies the production of collective capabilities to produce moral economies, so we could speak about *care economies*. A moral economy emerged out of disaster when people developed a collective interpretation of the quake and developed collective capabilities with the revitalization of traditional values (Bastide, 2016). In another way, for example in Indonesia, as a moral economy emerged *gotong royong* (mutual assistance) revitalized collective life in local communities: *gotong royong* is a fundamental identification "device" among Javanese which contains "ethnic" values of hard work, solidarity, and dedication to the community. Laura Centemeri (2015), in working on disaster memory, showed how residents, faced with orders from institutional actors, refused to express their emotions, an overexposure of the intimate in the public sphere, while memory becomes a political and identity resource. In China through a case study in Wang Village in Anhui, Zhang Jinjun (2018) showed that the core elements of collective memory, such as "bitterness," "toughness," "resentment," and "hatred," have played an important role in the peasants' environmental protest behavior.

If the construction of eco-anxiety is increasingly becoming a fundamental concept in sociology, especially in sociology of emotions, Paul Jobin (2025) conceived the idea of *regimes of toxicity* as different discourses and knowledge on toxicity; he considers that the dominant episteme on toxicity—or what he calls the dominant *regime of toxicity*—ignores or downplays the physical and psychological latent risks for individuals and considers that the time bomb provokes a *chemo-anxiety*: this author is focusing on the struggle of the recognition the sufferings by the victims and their mobilization over thirty years. Paul Jobin considers that chemo-anxiety is produced by three regimes of historicity (Hartog, 2015): the role of *memory* as the presence of the past, *attention* as the presence of the present, and *expectation* as the presence of the future.

## 4 Disasters' Biopolitics, Trust, and *Flexible Citizenship*

Political modernization was used to conceptualize the multiplicity of actors and multi-level governance, and to contribute to the understanding of environmental policymaking (Van Tatenhove, Arts, 2000; Van Tatenhove, 2003). In Europe, environmental policies have been driven by growing demands from civil society. Since 1960 the associative mobilizations and their interactions with the administrative and political networks have produced public action. Companies have been described as resistant to the green economy. Pierre Lascoumes (2018) has shown that the environment has taken its place in government agendas in Europe under the pressure of social movements and groups of experts; he emphasized the loss of the centrality of the state.

The status of controversies very quickly appeared central in the construction of public environmental problems, and the sociology of science became central to understanding the plurality of contradictory rationalities and assemblages of knowledge and perceptions. Two theoretical approaches have mainly been mobilized: the problematic of bounded rationality from Simon and Boudon's approach focuses on rationality based on "the right reasons to do it" for actors in situation. In European sociology in a broader perspective, the forming of emergent "public arenas" (Cefaï, 2007; Poulain, 2013) raises the issue of their democratic nature, as assessed by the modes of participation and the nature of democratic processes involved in the reconstruction period. The process of the emergence of "public arenas" and of different commitment regimes pertains to distinct institutional actors, their forms of mobilization, and their coordination to produce "interactional citizenship" (Colomy and Brown, 1996).

Resilience has become a central category in disaster policies (Giry, 2020) in both Europe and Asia. According to Benoît Giry (2023), we could distinguish post-disaster policies into the categories of planning policies and resilience policies: the former are based on a top-down, centralized model that denies victims' capacity for action, while the latter are organized around "interactional citizenship." Numerous studies have shown that the resilience process is built on social capital (bonding social capital and bridging social capital), human capital, material capital, and regenerative skills (Aldrich, 2012). Benoît Giry (2023) distinguishes between a conservative definition of resilience as the ability of a community to recover what "made society" before the disaster, and a second form of resilience, "autopoietic" and progressive, which develops through the acceptance of a world that is different from the one that existed before the disaster.

Japanese sociologists (Tsukahara, 2011) have demonstrated how the notion of reconstruction and resilience have been used in the policy of "creative reconstruction" and how, in a critical perspective, the reconstruction plan caused the destruction of human relationships, the social isolation, the precarization and the marginalization of some populations. In the theory of disaster capitalism Shiozaki (2010) called this phenomenon "reconstruction disaster." Through the four categories of defiance, flight, captivity, and expulsion (November, 2008, 2016) Rina Kojima (2023) analyzed the reconstruction policy's socio-cultural consequences on the victims of the Fukushima nuclear accident who left, stayed, or returned to the contaminated territories in a process of individual and collective resilience. Finally, for Shigeyoshi Tanaka (2023) the marginalization of disaster victims and disaster-afflicted communities appears as a characteristic of the reconstruction project.

In China Li Peilin and Wang Xiaoyi (2016), using the political definition of "ecological civilization construction," have shown that multi-governance is constructed around ecological migration organized by local and central governments. Jiang Pei (2019) did research on the issue of rural environmental governance in combining rural interest, karma, and kinship to expand rural social capital, and repair social relations within rural society in combining rural interest, karma, kinship, and rural social capital. Zhang Qian (2019) considered socio-economic systems, natural ecosystems, and local culture to analyze how institutional arrangements and ecological knowledge are tested and revised in the fabric of adaptive governance and climate change. Several surveys have highlighted the low level of environmental awareness; Bao Zhiming and Chen Zhangjiang (2011) explained this phenomenon as being due to a weak civic culture, and devices for the control and concealment of information on environmental problems. Chen Tao and Guo Xueping (2025) criticize governance theories, on the one hand by highlighting the gaps, distances, and ruptures between "disaster governance and farmers' lives," and on the other hand by considering farmers' autonomy, capabilities, and indigenous knowledge in rural environmental management.

Thinking about disasters' subpolitics also implies the production of collective capabilities by victims, moral economies, and care economies. Motohiko Nagata (2016) shows how survivors and volunteers collaborate for the community, creating an "endogenous revitalization." Different commitment regimes pertain to distinct institutional actors, their forms of mobilization, and their coordination to produce "flexible citizenship" (Ong, 1999) and forms of acknowledgment of human suffering. As a result, new and old forms of social coordination are embedded and dissociated in a diversity of institutional and non-institutional places, as individual and collective actors are faced with the huge task in hand.

*Geographies of care* are producing new social solidarities, diffracted inequalities, and new moral boundaries; they also sustain new forms of participative democracy. Institutional actors and citizens are producing different normative repertories and diverse regimes of action, like conciliation, negotiation, and dispute (Roulleau-Berger, 2016). As part of a sociology of attachments to think about disasters, the concept of *environmental care* was introduced; that is to say, practices of care brought to places and ways to keep the environment of the inhabitants (Centemeri, 2015). Faced with modes of government by unrest and an economy of fear in the context of catastrophes, Sandrine Revet (2013) proposed to be focused on ruses and the arts that citizens make with a government of risk.

In environmental and disaster sociology the concepts of knowledge, risk perception, acceptance, and public attitudes are still central. He Guangxi, Zhao Yandong, Zhang Wen-Xia, and Xue Pin (2016) demonstrated that the degree of people's acceptance of Genetically Modified Foods (GMF) is influenced by knowledge (government, experts (scientists), and mass media). These authors explain that, faced with a new technology like transgenic technology, individuals can only make a "bounded rational choice" due to their limited knowledge. Guo Yuhua (2004) has introduced a specific and very pertinent Chinese concept of "knowledge power" to define political and scientific authority and the interplay between new Internet-based media and traditional media. Employing "knowledge power" is a way of dealing with the social construction of trust and distrust. Chinese sociologists also dealt with environmental inequalities in emphasizing the issue of the reconstruction of trust, especially in post-disaster contexts; for them, distrust, more fundamentally, is one of the great challenges in Chinese society. Zhao Yandong and Shi Changhui (2016) distinguished:

- Interpersonal trust, including the trust in both acquaintances and strangers
- Institutional trust, including the trust in both governments and social institutions

These authors consider that institutional trust is the key to the maintenance of the social system and individuals' "ontological security." They demonstrate how interpersonal trust and institutional trust produce a social trust structure as a dynamic, unstable, and reversible process in which different kinds of trusts change before, during, and after the disaster. To understand this process, scholars propose an analysis of the trust among affected people, the trust between the affected people and the government, the trust between the government and NGOs, and the trust between affected people and NGOs. Communities with higher levels of trust will recover more rapidly and better, which means that the variations of degrees of trust will influence the process of re-creating society after a disaster. Social participation, the performance of the government and institutions, and norms and values, as well as personality, are defined as social determinants of trust. Chinese sociologists consider trust as a social and political process which influences normative orders and individual identities in risk or disaster societies (He, Shi, Zhang, Ma, Zhao, 2012; Shen, Sun, Su, 2012; Su, Zhang, Gao, Li, 2008).

## 5 Capabilities and Mobilizations

How can individual and collective capabilities either reinforce or compensate for inequalities depending on the choices made when facing a disaster? This signifies the mobilization of economic, social, symbolic, and cultural resources to produce regimes of action, for example, flight, defiance, and expulsion (November, 2008, 2016). Citizens facing risks and post-disaster consequences develop economies of political judgment by producing regimes of action from the responses of the institutions managing the post-disaster situation; these regimes of action of the victims mark out other moral boundaries linked to citizen spaces which arise out of silence, complaint, consent, indignation, or distancing with institutions (Roulleau-Berger, 2016). According to the times, places, and political and economic configurations, the citizens pass from one regime to another and do so according to the competences and regimes of action of the institutional actors.

In Japan (Nagata, 2012; Chen, Sugiman, 2010), each time there was an encounter with a huge disaster, citizens learned something new and adapted themselves to a similar disaster. Adaptive learning means resocialization and the production of new identities. Individuals have no choice but to be involved in the re-creation of society. The re-creation of society could mean maintaining previous forms of socialization, inventing new ones, breaking with previous ones, or finding compromises between previous and present social, economic, and moral forms. All social groups are unequal in the process of re-creating society. When confronted with different types of risks or disasters, individuals have to develop capabilities within a resocialization process. Luo Hongguang and Liu Zhenghai (2016) have defined different kinds of capabilities in post-disaster context: change of living space, change of living mode, restoration of traditional resources, cooperation with local government, reconstruction of trust, and the production of folk religious practices.

We can distinguish strong, intermediate, and weak survival capabilities in facing the disaster and evacuation; some victims who have developed strong survival capabilities tried to evacuate but failed, and other victims could not evacuate (Tanaka, 2023). While disaster victims develop individual capabilities, they also produce collective capabilities that are expressed in collective mobilizations and actions. In Western Europe, environmental movements emerged as early as 1960, inextricably linked to the process of industrialization and European colonial domination. Since the 1970s ecological mobilizations have continued to multiply, they are characterized today by a double movement of radicalization and institutionalization (Villalba, Ollitrault, 2012) which focuses on a "convergence of struggles" and the formation of coalitions between ecological movements (Zaccai, Orban, 2017). During the 2010s, research focused more on new forms of collective territorialization, such as the *Indignados* movement in Madrid and *Occupy Wall Street* in London, followed by climate protests and the emergence of new collectives acting to demand political autonomy or even civil disobedience. The diversity of action repertoires in environmental mobilizations has led to a rethinking of the question of societal transformations (Cary, Villalba, 2023). Since the 1970s, multiple forms of ecofeminism have also emerged, based on the one hand on struggles against various forms of domination, including male domination and the domination of nature by humans, and on the other hand on experimentation with lifestyles that herald a world free from capitalism (Comby and Dubuisson-Quellier, 2023). Isabelle Stengers (2019) asserted that in "the ecofeminism of the 1980s, women invented modes of action that rejected the devastation of the environment in struggles against what threatens life on Earth." In another way, Jérôme Michalon (2014) has studied situations relating to human/animal relationships, including the emergence of pro-animal mobilizations. The animal cause is a site where the question of political representation crosses that of scientific representation, which makes it a privileged point of observation of the relationship between science and activism. While in the United States the question of inequalities, even environmental racism and environmental justice, has been raised to analyze social movements and mobilizations, a decolonial critique of environmental justice remains more timid in France (Keucheyan, 2018).

Theoretical approaches also put more emphasis on what makes the "commons," others on the environmental inequalities which are deepened from below (reassembly of the poorest) but also from the top (more people have greater access to a preserved environment) (Larrère, 2017); this approach, based on "the commons" according to Anna Tsing (2017), remains Western-centric in that nature is considered to be the property of humans. It therefore gives rise to the concept of "latent commons" that arises from "biotic communities" that are not exclusively human (Cary, Villalba, 2023).

In post-disaster contexts, although individual subjectivities are traumatized, individuals could nevertheless be able to assemble individual and collective capabilities, and those of social and economic organization. New horizons of coexistence and coordination unfold and public arenas are opened by institutional actors, intermediate actors, and citizens. Citizens are involved in emerging processes of re-socialization and re-creation of "what makes society" in *intermediate spaces* or "tiers-lieux," for example, in France (Roulleau-Berger, 1991, 2003) or in some philosophy cafés, to hear the voices of the victims in Japan with *intermediate intellectual people* (Yama, 2020, 2022).

In China during the last ten years environmental protests and demonstrations (Sun, Zhao, 2007) have been the subject of studies focused on the analysis of theories of social movements to study the organization, mobilization, and strategy of various events from the theory of resource mobilization, political opportunities, and frameworks of action.

The work of Chen Tao and Xie Jiabiao (2016), starting from the question of the environmental resistance of the peasants, shows that the question of collective mobilization in China is built today through processes of influence, diffusion, and diffraction between distinct forms of action that can mobilize "weak actors" who oppose "strong actors," but especially new types of intermediate actors whose status and position vary in the different grouping configurations.

Chinese scholars now are paying more attention to the internal structure of environmental protests and the social mentality of the subaltern groups. In the Louis Island fishermen's environmental contention studied by Chen Tao and Yang Yue (2016), the mentality of protesters experienced four-stage change, namely "seeking social justice–imbalance–weak identity–rectification of name." Their mentality towards the local government varies from dependence, fear, and resentment to understanding. In Shagang Village, Luo Yajuan (2015) found that villagers' actions were all in the normative systems characterized by the differential mode of rite and norm: whether they protested the polluting enterprise or interacted with the government through petitions. In the suburban areas of China, the farmers' environmental resistance bears the characteristics of contingency which is different from the Western large-scale environmental movement (Gao, Xu, 2015). Zhang Jieying (2019) introduced the issue of waste governance and green activism in China in showing how a 'civic' governing space is expanded by intermediate actors under the radar of state regulation. Through a case study in Wang Village in Anhui, Zhang Jinjun (2018) showed that the core elements of collective memory, such as "bitterness," "toughness," "resentment," and "hatred" have played an important role in the peasants' environmental protest behavior.

For Feng Shizeng (2023), in contemporary China, the presumed collective action and social movements tend to be labeled by the official as a "group event." The author traces the historical change of the state's conception of "group event" from 1949 to show that whereas China's regime has transitioned from an ideocracy to a technocracy, the state's conception of social conflicts undergoes a reverse change from technical to political. He offers a post-Western sociological perspective by highlighting a crucial element understudied in the Eurocentric tradition: state-targeting social movements in interaction with the technocratization of a post-revolutionary regime.

After the Fukushima accident, anxiety, distrust, and "protest in the street" characterized Japan's civil society. Koichi Hasegawa (2004, 2021) discussed the historical outline of social movements, especially environmental movements, focusing on four major topics in four stages of the movements: 1) political movements of class struggle centered around labor movements by the socialist and communist party in 1945–1965; 2) citizens' movements and residents' movements in 1965–1975; 3) the end of the conservative-progressive ideological framework, especially in 1989; and 4) the collapse of the Japanese-style system—institutionalization of movements, and diversifying activities, including environmental NGOs and other NPOs. Koichi Hasegawa (2015) introduced in the sociology of social movements the problematic of continuities and discontinuities of political activism after the Fukushima disaster. On the one hand, he analyzed which resources can be mobilized and under what conditions; on the other hand, he demonstrated how interests, values, and beliefs of citizens and activities of social movements can be adjusted and brought into alignment—in the sense of the analytical framework from Benford and Snow (2000). Chigaya Kinoshita (2023) explained the anti-nuclear movement of the 2010s in Japan was due to their rebellion against the old developmental system of governance as well as against the old anti-system movement and for renewal in Japanese society's ongoing structural crisis. After 2011 students' mobilizations were characterized by interaction and solidarity with young people from Hong Kong, Taiwan, and Okinawa. So, the social movement of the 2010s in Japan is inserted in the global geography of social mobilizations.

## 6 Conclusion

Social studies of disasters in East Asia or Europe exacerbate what is essential to sociology, a global sociology that transcends Western hegemony and imposes a theoretical and epistemological shift in order to understand "what makes society" in the context of ecological crisis. The rise of the sociology of emotions, which has become part of the sociology of disasters, reveals blind spots in the various theoretical traditions that we are invited to identify in order to revisit concepts such as subjectivation, subjectivity, agency, capabilities, etc. The limits of political sociology, which allow us to understand the different "governments of disasters" (Revet, 2013) also become apparent but do not provide access to the individual and collective subpolitical narratives produced by disaster victims. Not only does the production of sociological knowledge about disasters require non-Western thinking, the practice of *methodological cosmopolitanism* (Beck, 2006), and a return to the role of the researcher in a global public space, in order to advance the invention of a *post-Western sociology* (Roulleau-Berger, Li, Kim, Yazawa, 2023).

## Bibliography

Aldrich, D.P. (2012). *Building resilience. Social capital in post-disaster recovery*, Chicago: University of Chicago.

Bao, Z.M., Chen, Z.J. (2011). "The Environmental Dimension of the Chinese Experience: Its Scope and Limits—A Review and Reflection on Environmental Sociology Research in China (Zhongguo Jingyan de Huanjing Zhiwei: Xiangdu Jiqi Xiandu—Dui Zhonguo Huanjing Shehuixue de Huigu yu Fansi)." *Sociological Studies (Shehuixue Yanjiu)*, 26 (6), 196–210, 245. doi:10.19934/j.cnki.shxyj.2011.06.009.

Bao, Z., & Zeng, W. (2021). Livelihood transformation and ecological environmental change: A case study of Village Y in Yunnan Province (Shengji zhuanxing yu shengtai huanjing bianqian—jiyu Yunnansheng Y cun de ge'an yanjiu). *Social Sciences in Yunnan (Yunnan Shehui Kexue)*, (2), 158–164.

Bao, Z.M., Yan, Q.S. (2022). "Population Mobility, Information Media, and Public Environmental Knowledge (Renkou Liudong, Xinxi Meijie Yu Gongzhong Huanjing Zhishi)." *Social Development Studies (Shehui Fazhan Yanjiu)*, 9 (2), 193–208, 246.

Barbot, J., Dodier, N. (2010). "Violence et démocratie au sein d'un collectif de victimes. Les rigueurs de l'entraidede." *Genèses*, 4 (81): 84–103.

Bastide, L. (2016). "The moral side of disaster: religion and post-quake recognition regimes in Java," in Li, P., Roulleau-Berger, L. (eds): *Ecological risks and disasters—New experiences in China and Europe*, London & New York: Routledge Publishers, pp. 111–122.

Beck, U. (2006). *The Cosmopolitan Vision*. Cambridge: Polity Press.

Beck, U. (2013). "Risk, class, crisis, hazards and cosmopolitan solidarity/risk community—conceptual and methodological clarifications." n°31, April 2013, Working Papers Series, FMSH.

Beck, U., Grande, E. (2010). "Varieties of second modernity: the cosmopolitan turn in social and political theory and research." *The British Journal of Sociology*, Volume 61, Issue 3, pp. 598–620.

Benford, R.D., and Snow, D.A. (2000). "Framing Processes and Social Movements: An Overview and Assessment." *Annual Review of Sociology*, 26, pp. 611–639.

Bullard, R.D. (1993). *Confronting environmental racism: voices from the grassroots*. Boston: South End Press.

Cary, Villalba (dir). (2023). *Sociologie de l'environnement*, Neuilly: Atlante.

Cefaï, D. (2007). *Pourquoi se mobilise-t-on? Les théories de l'action collective*, Paris: La Découverte.

Centemeri, L. (2013). "Enquêter la mémoire discrète du désastre de Seveso," in Revet, L., Langumier, J. (eds) *Le gouvernement des catastrophes*, Paris: Khartala.

Centemeri, L. (2015). "L'apport d'une sociologie des attachements pour penser la catastrophe environnementale." *Raison Publique*, 2015. https://hal.science/hal-01163221v2.

Chang, K.S. (2010). "The second modern condition? Compressed modernity as internalized reflexive cosmopolitization." *The British Journal of Sociology*, Volume 61, Issue 3.

Chateauraynaud, F., Torny, D. (1999). *Les Sombres Précurseurs. Une sociologie pragmatique de l'alerte et du risque*, Paris: Éditions de l'EHESS.

Chen, A.J. (2015). "The Origins and Development of Environmental Sociology (Huanjing Shehuixue De Youlai Yu Fazhan)." *Journal of Hohai University* (*Philosophy and Social Sciences Edition*) (*Hehai Daxue Xuebao, Zhexue Shehuikexue Ban*), 17 (5), 32–40, 104.

Chen, J. (2016). "'惹弥' 与' 惹撇' :纳人达巴仪式中的时空认知逻辑," in 民族研究, Vol. 4.

Chen, T., Guo, X. (2025). Nèi shēngxìng guānlián yǔ xiāngcūn huánjìng zhìlǐ zhōng nóngmín zhǔtǐ xìng de jiàngòu—jīyú lín jiāng xiàn "hédào zì guǎn" de ànlì yánjiū [Endogenous correlation and the construction of farmers' subjectivity in rural environmental governance: a case study based on "river self-management" in Linjiang County, Shèhuì xué yánjiū"] 2025 n°2, *Sociological Research*, Issue 2, 2025.

Chen, T., Xie, J. (2016). "The mixed resistance: Analysis of environmental struggle by peasants." *Sociological Studies* 3: 25–46.

Chen, T., & Yang, Y. (2016). The social psychology of grassroots groups in environmental contention: A case-based interpretation and extended analysis of the Penglai 19-3 oil spill incident (Huanjing kangzheng zhong diceng qunti de shehui xinli—jiyu Penglai 19-3 yiyou shijian de anli chanshi yu tuozhan fenxi). Journal of China University of Geosciences (Social Sciences Edition) (Zhongguo Dizhi Daxue Xuebao (Shehui Kexue Ban)), 16 (1), 102–112.

Chen, Y., Sugiman, T. (2010). "Relief activities of a Chinese non-governmental organization for victims of the Sichuan huge earthquake in 2008." *Japanese Journal of Group Dynamics*, 27, 131–157.

Colomy, P.J., Brown, D., 1996. Goffman and interactional citizenship, Sociological Perspectives 39 (3): 371.

Comby and Dubuisson-Quellier, (dir). (2023). *Mobilisations écologiques*, Paris, PUF.

Eisenstadt, S. (2000). *Multiple modernities*. New Brunswick: Transaction Publishers.

Faburel G. (2016). "The environment as a factor of spatial injustice: a new challenge for sustainable development of European regions?" *Sustainable Development Handbook—Policy and Urban Development—Tourism, Life Science, Management and Environment*, Intech Online Ed., Elsevier, 431–478.

Faburel, G. (2020). *Pour en finir avec les grandes villes*. Paris: Le Passager clandestin.

Feng, S. (2023). "State's temperament and the Control of Collective Action in Contemporary China," in Roulleau-Berger, L., Li, P., Kim, S.K., Yazawa, S. (eds): *Handbook of Post-Western Sociology. From East Asia to Europe,* Leiden & Boston: Brill Publishers, pp. 719–731.

Fischer, C. (2003). *L'homnivore : le goût, la cuisine et le corps.* Paris : Odile Jacob.

Gao, X., & Xu, J. (2015). Contingent countermovement: A sociological interpretation of peasants' environmental contention in China's social transition—A case study of

Village Q (Quanbian shi fanxiang yundong: shehui zhuanxing qi nongmin huanjing kangzheng de shehuixue chanshi—Q cun ge'an yanjiu). Journal of Hohai University (Philosophy and Social Sciences) (Hehai Daxue Xuebao (Zhexue Shehui Kexue Ban)), 17 (3), 39–44.

Gilbert, C., Henry, E. (2012). "Defining social problems: tensions between discreet compromise and publicity." *Revue française de sociologie*, English issue vol. 53, 31–54.

Giry, B. (2020). "Résilience territoriale," in Cole, S., Guignier, S. and Pasquier, R. *Dictionnaire des politiques territoriales*, Paris: Presses de Science Po, pp. 482–487.

Giry, B. (2023). *Sociologie des catastrophes*, Paris: La Découverte.

Guo, Y. (2004). "Angel or Devil: Social and Cultural Perspective to GM Soybean in China." *Journal of Huazhong University of Science and Technology*, n°12.

Hartog, F. (2015). *Regimes of Historicity: Presentism and Experiences of Time*. New York: Columbia University Press.

Hasegawa, K. (2004). *Constructing Civil Society in Japan, Voices of Environmental Sociology*, Melbourne: Transpacific Press.

Hasegawa, K. (2015). *Beyond Fukushima: Toward a Post-Nuclear Society*, Melbourne: Transpacific Press.

Hasegawa, K. (2021). "Japanese Environmental Sociology: Focus and Issues in Three Stages of Development." *International Sociology Reviews* 36 (2): 289–301.

Hasegawa, K. (2023). *Contributions of Japanese environmental sociology*, in Roulleau-Berger, L., Li, P., Kim, S.K., Yazawa, S. (eds.), *Handbook of Post-Western Sociology. From East Asia to Europe*, Leiden & Boston: Brill Publishers, pp. 691–710.

He, Guangxi, Changhui Shi, Wenxia Zhang, Ying Ma, and Yandong Zhao. (2012). "Monitoring Society Reconstruction and Evaluating Residents' Lives in Wenchuan Earthquake-affected Area (2008–2011)." In Xin Ru, Xueyi Lu, and Peilin Li (eds.), Blue Book of China's Society: Society of China: Analysis and Forecast (2012), 195–212. Beijing: Social Sciences Academic Press (China).

He, G., Zhao, Y., Zhang Wen, X., Xue, P. (eds). (2016). "Knowledge, Trust, Risk Perception, and Acceptance: A Sociological Analysis on Public Acceptance of GM Crops in China," Li, P., Roulleau-Berger, L. (eds): *Ecological risks and disasters—New experiences in China and Europe*, London & New York: Routledge Publishers pp. 58–76.

Hochedez, C., Le Gall, J. (2016). Justice alimentaire et agriculture: introduction. *Justice spatiale = Spatial justice*, 2016, Justice alimentaire et agriculture, 9. ⟨hal-01342994⟩.

Hong, D., Ku, D. (2023). "The Legacy of the Development State and the Rise of Fragmented Green Growth," in Roulleau-Berger, L., Li, P., Kim, S.K., Yazawa, S. (eds.): *Handbook of Post-Western Sociology. From East Asia to Europe,* Leiden & Boston: Brill Publishers, pp. 812–823.

Hong, D.Y. (2017). "Environmental Sociology: Facts, Theories, and Values (Huanjing Shehuixue: Shishi, Lilun Yu Jiazhi)." *Thinking Front (Sixiang Zhanxian)*, 43 (1), 78–92.

Hong, D.Y., Lu, C.T. (2011). "A Multi-Level Analysis of Public Environmental Concern: Based on CGSS2003 Data from China (Gongzhong Huanjing Guanxin De

Duoceng Fenxi)." *Sociological Studies (Shehuixue Yanjiu)* 26 (6), 154–170, 244–245. doi:10.19934/j.cnki.shxyj.2011.06.007.

Iijima, K., ed, (1993). *Kankyo Shakaigaku* (environmental sociology), Tokyo: Yuhikaku.

Jiang, P. (2019). The social foundation of rural environmental governance: A case study based on the classification of garbage in Lujia Village, Zhejiang Province Workshop "Compressed Modernities, Ecological Risks and Disasters in Europe, Asia, Latin America, Africa", LIA CNRS/ENS Lyon–CASS, June Lyon, France.

Jobin, P. (2025). The Post-Western Anthropocène, in Roulleau-Berger, L., Li, P., Kim, S.K., Yazawa, S. (eds.): Handbook of Post-Western Sociology. From East Asia to Europe, Leiden & Boston: Brill Publishers, pp. 786–799.

Jobin, P. (2025): "Chemo-anxiety and the regimes of toxicity." *BioSocieties,* https://doi.org/10.1057/s41292-025-00357-2.

Keiichi, S., Yamamoto, H. (2023). "Introduction: Japanese experience of great disasters: Revisioning 3/11 and beyond." *Japanese Journal of Sociology*, 3–6.

Keucheyan, R. (2018). *La Nature est un champ de bataille.* Essai d'écologie politique, Paris: La Découverte.

Kinoshita, C. (2023). "Historical meaning of post-Fukushima social movements: crisis in hegemony in post-war Japan and prefigurative politics in 2010s." *Japanese Journal of Sociology*, 32, 56–68.

Kojima, R. (2023). "Reconstruction and resilience after Fukushima: a critical analysis of nuclear risk and disaster." *Japanese Journal of Sociology*, 32, 25–44.

Larrère, C. (dir.). (2017). *Les inégalités environnementales*, Paris: PUF.

Lascoumes, P. (2018). *Action publique et environnement*, Paris: PUF.

Li, P., Roulleau-Berger, L. (eds), (2016). *Ecological risks and disasters—New experiences in China and Europe*, London & New York: Routledge Publishers.

Li, P., Wang, X. (2016). "Reducing Double Risks in Ecological Degradation and Poverty—A Research on Ecological Migration in Ningxia Hui Autonomous Region of China," in Li, P., Roulleau-Berger, L. (eds): *Ecological risks and disasters—New experiences in China and Europe*, London & New York: Routledge Publishers, pp. 122–135.

Liu, Z. (2009). "The Cult of Ancestor Worship and Inheritance of Folk Culture," in *Anthropology of Religion*, 170–196. (Chinese version.)

Luo, H., Liu, Z. (2016). "A study on the internal generation mechanism of the post-disaster reconstruction" in Li, P., Roulleau-Berger, L. (eds.): *Ecological risks and disasters—New experiences in China and Europe*, London & New York: Routledge Publishers, pp. 24–41.

Luo, Y. (2015). Differential order and ritual propriety: A structural analysis and local-meaning interpretation of peasants' environmental contention—A case study of Shagang Village (Chaxu liyi: nongmin huanjing kangzheng xingdong de jiegou fenxi ji xiangtu yiyi jiedu—Shagangcun ge'an yanjiu). Journal of China Agricultural University (Social Sciences Edition) (Zhongguo Nongye Daxue Xuebao (Shehui Kexue Ban)), 32 (4), 59–67.

Michalon, J., (2014). Panser avec les animaux. Sociologie du soin par le contact animalier, Presses des Mines, Paris.

Mol, A.P.J., Sonnenfeld, D.A. (2000). *Ecological Modernization Around the World. Perspectives and Critical Debates*. London and Portland: Frank Cass/Routledge.

Mol, A.P.J., Spaargaren, G. (2000). "Ecological Modernization and Consumption: A Reply." *Society and Natural Resources*, 17, 261–265.

Nagata, M. (2012). "A 'soft' volunteerism in super-extensive disaster: case of Noda," in Shaw, R., Takeuchi, Y. (eds.): *East Japan Earthquake and Tsunami Evacuation, Communication, Education and Volunteerism*. Research Publishing Services.

Nagata, M. (2016). "A bottom-up counterpart assistance for revitalization of community after Great East Japan Earthquake: A case of a volunteer network 'Team North Rias' supporting Noda village" in Li, P., Roulleau-Berger, L. (eds.): *Ecological risks and disasters—New experiences in China and Europe*, London & New York: Routledge Publishers, pp. 157–170.

Niyama, Y. (2012). The public psychology and an interactive risk communication on effects of radiation contamination of food on health : scrutinizing information necessary for acquisition of knowledge, *Journal of Rural Problems*, 48 (3): 346–354.

Nomiya, D. (2023). "Voice of the Dead: Hibakusha Collective Memory against the Western Ethos," in Roulleau-Berger, L., Li, P., Kim, S.K., Yazawa, S. (eds.), 2023. *Handbook of Post-Western Sociology. From East Asia to Europe*, Leiden & Boston: Brill Publishers, pp. 292–306.

November, V. (2008). "Spatiality of risks." *Environment and Planning A*. 40, 1523–1527.

November, V. (2016). "Risks, planning and socio-spatio-temporal differentiation," in Li, P., Roulleau-Berger, L. (eds.): *Ecological risks and disasters—New experiences in China and Europe*, London & New York: Routledge Publishers, pp. 43–58.

Nygren, A. (2014). "Eco-Imperialism and environmental justice," in Lockie, S., Sonnenfeld, D.A., Fisher, D.R. (2014), *Routledge International Handbook of Social and Environmental Change*, London & New York: Routledge, pp. 58–70.

Ong, A. (1999). *Flexible Citizenship: The Cultural Logics of Transnationality*, Durham: Duke.

Pollak, M. (1998). *Les homosexuels et le sida. Sociologie d'une épidémie*, Paris: Métailié.

Poulain, J.P. (2013), *Sociologies de l'alimentation: les mangeurs de l'espace social alimentaire*, Paris: PUF.

Poulain, J.P. (2019). *Sociologies de l'alimentation*, Paris: PUF.

Revet, L., Langumier, J. (2013). *Le gouvernement des catastrophes*, Paris: Karthala.

Roulleau-Berger, L. (1991). *La Ville-Intervalle, Jeunes entre centre et banlieue*, Paris: Méridiens Klincksieck.

Roulleau-Berger, L. (2003). "Economie solidaire, non-reconnaissance publique et production d'espaces intermédiaires," in *Economie solidaire et démocratie, Revue Hermès*, deuxième trimestre.

Roulleau-Berger, L. (2016). "Life exposed, inequalities and moral economies in post-disaster societies: China, Japan, Indonesia," in Li, P., Roulleau-Berger, L. (eds.): *Ecological risks and disasters—New experiences in China and Europe*, London & New York: Routledge Publishers, pp. 135–155.

Roulleau-Berger, L., Li, P., Kim, S.K., Yazawa, S. (eds). (2023). *Handbook of Post-Western Sociology. From East Asia to Europe,* Leiden & Boston: Brill Publishers.

Shen, H., Sun, X., Su, J. (2012). "Trust in Science and Technology, Trust in study of natural disaster." *Sociological Studies*, (5), 164–245.

Shiozaki, Y. (2010). "Daishinsai no 15 Nen (15 Years after the Great Earthquake)," in Shiozaki, Y., Nishikawa, E. (eds.), *Daiṣhinsai 15 Nen to fukkô no Sonae* (15 Years after the Great Earthquake and Preparations for Reconstruction), Kurieitsu Kamogawa: Shun'ichi Deguchi and Hyôgoken shinsai fukkô kenkyû sentâ. Kyoto, pp. 8–15.

Stengers, I. (2019). *Résister au désastre*, Paris: Editions Wild project.

Su, J., Zhang, M., Gao, L., Li, N. (2008). "Impact of trust in Flood prevention projects on public risk perception: an analysis based on questionnaire in Yangtze River Basin." *Journal of Natural Disasters*, 17 (1), 75–80.

Sun, Y., & Zhao, D. (2007). Multifaceted state and fragmented society: Dynamics of environmental movement in China. In D.L. Yang (Ed.), Discontented miracle: Growth, conflict, and institutional adaptations in China (pp. 111–159). World Scientific.

Tanaka, S. (2023). "The 2011 Great East Japan Earthquake and Tsunami: The Highest Casualties and Largest Reconstruction Funds—Characteristics of Major Disasters and Future Challenges in Developed Countries." *Japanese Journal of Sociology* 32 (1): 7–24. https://doi.org/10.1111/ijjs.12147.

Therborn, G. (2003). "Entangled Modernities." *European Journal of Social Theory* 6 (3): 293–305. https://doi.org/10.1177/13684310030063002.

Tierney, K. (2019) *Disasters. A sociological approach*, Cambridge: Polity Press.

Torigoe, H., Kada, Y. (eds.). (1984). *Mizu to Hito no Kankyoshi* (Environmental History of Water and People. The Lake Biwa report). Tokyo: Ochanomizu Shobo.

Tsing, A. (2017). *Le champignon de la fin du monde*, Paris: La Découverte.

Tsukahara, T. (2011). *Saigai shihon shugi no hatsudô—nido hakai sareta kôbe kara nani wo manabu no ka*? (Launching Disaster Capitalism—What Can We Learn from Twice Destroyed Kobe?). *Gendai Shisô* 39 (7): 202–211.

Van Tatenhove, J., Arts, B., Leroy, P. (eds.). (2000). *Political Modernization and the Environment: The Renewal of Policy Arrangements*, Dordrecht: Kluwer.

Van Tatenhove, J., Pieter, L. (2003). "Environment and Participation in a Context of Political Modernization." *Environmental Values*, 12, 2, pp. 155–174.

Vásquez Lezama, P. (1999). "De la tragédie collective à l'individuation du malheur: L'expérience de fin de légitimité de la condition de victime des sinistrés de la catastrophe La Tragedia (1999) au Venezuela," *Anthropologie et Sociétés*, ID: 10.7202/039688ar.

Vergès, F. (2017). "Racial Capitalocène: is the Anthropocène racial?" in Johnson, G.T., Lubin, A. (eds.), *Future of black radicalism*, London & New York: Verso, pp. 72–82.

Villalba, B., Ollitrault, O. (2012). "Sous les pavés, la Terre. Mobilisations environnementales en France (1960–2011): entre contestations et expertises," in Danielle Tartakowsky and Michel Pigenet (dir.), *Histoire des mouvements sociaux en France*, Paris: La Découverte, pp. 716–723. ⟨hal-03081479⟩.

Walker, G. (2009). "Beyond Distribution and Proximity: Exploring the Multiple Spatialities of Environmental Justice." *Antipode*, 41, 4, pp. 614–636.

Yama, Y. (2006).「『象徴的復興』とは何か」What is Symbolic Recovery? *Sentan Shakai Kenkyu*, 5.

Yama, Y. (2020).「『媒介的知識人』とは何か」[What is an "Intermediate Intellectual"?]. *Disaster Recovery Studies*, 11.

Yama, Y. (2022).「語り合う場のデザイン—哲学カフェの試みから」[Designing Dialogical Spaces: The Philosophy Café Approach]. *DPRI Newsletter*, 101, 7. Kyoto: Kyoto University Disaster Prevention Research Institute.

Yama, Y. (2023).「関東地震の災害遺構—慰霊・追悼と記憶の継承」[Disaster Heritage of the Great Kanto Earthquake: Memorials, Mourning, and the Transmission of Memory]. *FUKKOU*, 52, 6. Nishinomiya: Kwansei Gakuin University Institute for Disaster Recovery System Research.

Yamaguchi, M. (2019). Risk in Egg Production and Marketing in Japan, in Augustin-Jean, L., Poulain, J.P: Risk And Food safety in China and Japan. Theoritical Perspectives and Empiricla Insights, New York: Routledge.

Yazawa, S. (2014). "Civilizational Encounter, Cultural Translation and Social Reflexivity: a note on History of Sociology in Japan," in *A quest of East Asian Sociologies*, eds. Kim, S.K., Li, P., Yazawa, S.,. Seoul: Seoul National University Press, pp. 131–167.

Zaccai, E., and Orban, A. (2017). "Mobilisations écologiques actuelles, mobilisations des années 1960–1970: quels parallèles?" *Développement durable et territoires* [online], Vol. 8, n°2 | July 2017, DOI: https://doi.org/10.4000/developpementdurable.11847.

Zhang, J. (2018). Collective memory and peasants' environmental contention: A case study of Wang Village in Anhui (Jiti jiyi yu nongmin de huanjing kangzheng—yi Anhui Wangcun wei li). Journal of Anhui Normal University (Hum. & Soc. Sci.) (Anhui Shifan Daxue Xuebao (Renwen Shehui Kexue Ban)), 46 (1), 77–85.

Zhang, J. (2019). Waste crisis in contemporary China: Green governance, social activisms and technological controversies Workshop "Compressed Modernities, Ecological Risks and Disasters in Europe, Asia, Latin America, Africa", June, LIA CNRS/ENS Lyon–CASS, Lyon, France.

Zhang, Q., and Barrios, R.E. (2017). "Imagining culture: The politics of culturally sensitive reconstruction and resilience building in post-Wenchuan Earthquake China," in M. Companion, M., Chaiken, M. (eds.), *Responses to Disasters and Climate Change: Understanding Vulnerability and Fostering Resilience*, pp. 93–102. CRC Press.

Zhang, Y.L. (2006). "On the Marginalization of Rural Teachers in Contemporary China. (Guanyu Dangdai Zhongguo Xiangcun Jiaoshi de Bianyuanhua Wenti)." *Journal of South China Normal University (Social Science Edition) (Huanan Shifan Daxue Xuebao, Shehuikexue Ban)*, (1), 105–107.

Zhang, Y.L. (2019). "Urbanization and Ecological Environmental Issues in China: A Review of the 2018 China Humanities and Social Sciences Environmental Forum (Zhongguo De Chengshihua Yu Shengtai Huanjing Wenti—2018 Zhongguo Renwen Shehuikexue Huanjing Luntan Yanjiu Shuping)." *Journal of Nanjing Tech University (Social Science Edition) (Nanjing Gongye Daxue Xuebao-Shehuikexue Ban)*, 18 (1), 1–10, 111.

Zhang, Y.L. (2021). "The Yangtze River in the Age of the Anthropocene: The Social Dynamics of Life Depletion (Renleishi Shidai De Changjiang: Shengming Shuaijian De Shehui Dongli)." *Journal of Yunnan University (Social Sciences Edition) (Yunnan Daxue Xuebao-Shehuikexue Ban)*, 20 (3), 107–122. doi:10.19833/j.cnki.jyu.2021.03.013.

Zhao, Y. (2007). "Social Capital and Post-disaster Recovery: A Sociologic Study on Natural Disasters." *Sociological Studies*, Vol. 5.

Zhao, Y., Shi, C., (2016) The structure and change of social trust during post-disaster reconstruction : an example of the Wenchuan earthquake-affected population in Li, P., Roulleau-Berger, L. (eds), Ecological risks and disasters—New experiences in China and Europe, London & New York: Routledge Publishers.

CHAPTER 3

# Learning from the Disaster and the Post-disaster Recovery: Dreams and Reality 14 Years Later of the 2011 Great East Japan Earthquake

*Koichi Hasegawa*

## 1 Introduction: Dreams and Reality 14 Years Later of the 2011 Disaster

Understanding post-disaster recovery from a sociological perspective is of paramount importance. This approach allows us to grasp the full extent of the impact on community leaders, ordinary citizens, scientists, and sociologists. In this paper, I aim to delve into several key lessons, with a specific focus on earthquake and tsunami disasters. For a detailed analysis of the Fukushima nuclear accident and its social implications, see Hasegawa's work (2014, 2015, and 2018).

Following the Great East Japan Earthquake, many people have envisioned it as an opportunity to revitalize Japanese society and local communities. They also witnessed the scorched disaster areas reminiscent of battlefields. They were forced to endure the aftershocks that continued day after day. The Great East Japan Earthquake marked a turning point for Japanese society, presenting a choice between a real, sustainable future and continuing on the existing path (Samuels, 2013). What was the reality 14 years later in the tsunami-affected areas? Most of the hardware restoration and reconstruction projects have been completed. When you visit disaster-stricken areas, you will see many new buildings that look like plastic models and inorganic spaces that look like paperbacks. The warmth, rustle, and bustle have disappeared for a long time except for tourists. However, amidst this, the resilience and determination of the Japanese people shine through, inspiring hope for a better future.

Disaster victims lost their close family members, relatives, and friends, lost their homes, lost their hometowns, and lost their livelihoods. What was washed away by the tsunami cannot be expressed in numbers, such as the number of dead and missing people or the number of destroyed houses. Compared to the magnitude of the sacrifices, the victims, evacuees, and local communities have gained far too little. In many disaster-affected areas, residents' hopes for "sustainable reconstruction of their towns and lives" continued to be disappointed.

 HTTPS://DOI.ORG/10.1163/9789004751804-005

Delays in solving problems due to various vested interests, such as the construction industry's influence on policy decisions, knee-jerk responses, weak sense of crisis on the part of policymakers, and bureaucratic inflexibility in existing laws, are common structural features of many social and political problems in contemporary Japanese society. These were the fundamental factors behind the delay in reform that led to the 'lost 30 years.' The crisis was also an opportunity. However, Japanese society missed the best chance to transform policy structures and decision-making processes after the 2011 disaster. Can we find the exit or the way to transform these structures? The first lesson to be learned from the Great East Japan Earthquake is why Japan did not change after it and why it missed the chance to transform into a real, sustainable future.

## 2 Disasters Can Occur at Any time and Anywhere in Japan

The 2024 Noto Earthquake, which hit on January 1, shocked the Japanese. January 1 is the most sacred holiday in Japan. People were forced to realize that 1) earthquakes can occur anytime, anywhere in the Japanese archipelago, and 2) it is crucial to remember the lessons from the Great East Japan Earthquake and other massive past earthquakes and to prepare for the coming disasters. A trend that has become even more pronounced over the past 14 years is that "disasters have become commonplace." Earthquakes with a maximum seismic intensity of 7 have occurred six times in these 30 years since the 1995 Great Hanshin-Awaji Earthquake. Earthquakes with a maximum seismic intensity of 6 have occurred almost every year. Due to climate change, flood damage caused by localized torrential rains and typhoons has occurred almost every year over the past few years. The COVID-19 pandemic, which began in February 2020, is also a new type of disaster. Disasters and regular times are continuous, and disasters can occur at any time and in any place. "A disaster can be seen as a "snapshot" of a society that can be used to study the society itself." Disasters should be regarded "as continuous processes occurring from the pre-diaster period through to the reconstruction period, as well as part of social structure and change" (Tanaka, 2023, p. 23).

Examining natural disasters worldwide since 2000, we are acutely aware that many have occurred in Asia and the Pacific Rim region. Among developed countries, Japan is one of the most disaster-prone countries in the world, experiencing a range of natural hazards, including earthquakes, tsunamis, volcanic eruptions, and typhoon damage. The United Nations World Conference on Disaster Risk Reduction, held three times to date, has been hosted in Japanese cities: Yokohama in 1994, Kobe in 2005, and Sendai in 2015. The fact that frameworks named after these cities, such as *the Sendai Framework for Disaster Risk Reduction 2015–2030*, have been adopted each time demonstrates that Japan is one of the world's top disaster-prone countries.

On disaster prevention, mitigation, relief, recovery, or reconstruction, the Japanese government, local governments, and researchers have much to share with the world.

An author referred to the period following the Great East Japan Earthquake as the 'post-disaster' period. However, this period can be immediately seen as 'the period before new major disasters,' such as the Nankai Trough mega-earthquake and the direct earthquake in the metropolitan area. The period from 1918 to 1939, from the end of World War I until the outbreak of World War II with Nazi Germany's invasion of Poland, is often referred to as the 'interwar period' in Europe. Based on that, the current period should be called the 'inter-disaster period,' a term that emphasizes the ongoing nature of disaster management and preparedness rather than the post-disaster period.

Adopting the precautionary principle, it's clear that ongoing research is crucial for disaster prevention, mitigation, relief, recovery, and reconstruction. The second lesson from the Great East Japan Earthquake is the need for continuous preparedness and vigilance. We must be ready for the next major disaster, which can strike at any time and in any place. To prepare effectively, we must learn from the lessons of the past and share the areas that need improvement with society. By doing so, we can equip ourselves with the knowledge and understanding necessary to respond effectively to future crises, ensuring the safety and well-being of our communities.

## 3 Establishing a Permanent Ministry of Disaster Prevention and Reconstruction

The third lesson underscores the pressing need to establish a permanent Ministry of Disaster Prevention and Reconstruction. In Japan, the Self-Defense Forces are expected to provide disaster relief services in the event of a significant disaster alongside the fire department and police. However, Japan lacks a specialized organization comparable to the United States Federal Emergency Management Agency (FEMA), which is crucial for a disaster-prone country like Japan. Therefore, there should be a permanent ministry responsible for disaster relief, recovery, and reconstruction. The current temporary solution, the Reconstruction Agency, is only fixed until 2021 and prolonged until 2031. Due to its limited time, it will have no choice but to become a gathering place for relevant ministries and agencies, including the Ministry of Land, Infrastructure, Transport, and Tourism. The Reconstruction Agency's target areas are those affected by the Great East Japan Earthquake and the Fukushima nuclear accident and do not include areas affected by the 2016 Kumamoto earthquake or the 2024 Noto Peninsula earthquake. What the Noto Peninsula earthquake revealed is that the lessons learned from the Great East Japan Earthquake, such as how primary evacuation shelters should respond, have not been effectively passed down organizationally or systematically. This is particularly true at the national, prefectural, or municipal levels. A systematic mechanism is needed to pass on experience gained from recovery and reconstruction. It is necessary to establish a permanent organization specialized in the unified task of policy planning and coordination, from pre-disaster prevention and mitigation to recovery and reconstruction, to accumulate and disseminate knowledge and expertise, and to train specialized staff. The advantage of a permanently specified Ministry of Disaster and Reconstruction is that, like FEMA, it can provide one-stop command over fire and police departments and local governments in an emergency disaster, thereby significantly improving public safety and welfare.

The biggest problem with the Self-Defense Forces is that their primary mission is to defend the country from invading countries, and disaster relief is a secondary mission. As a result, the Self-Defense Forces do not have any equipment specifically designed for disaster relief. Disaster relief requires bulldozers and jeeps that can be easily and flexibly maneuvered, but the Self-Defense Forces only have large-scale equipment. The strength of the Self-Defense Forces lies in the size of their workforce. Based on military expertise and experience, they excel in aerial mobility, off-road mobility, communications, mass transportation, outdoor emergency medical care, and various logistical support capabilities. However, their weaknesses lie in the fundamental differences between military organizations and disaster relief organizations. Responding to major disasters with personnel and equipment is not necessarily easy. As a military organization, they are not suited to international disaster relief. It is not easy to work with volunteer organizations. They are chronically understaffed, with the occupancy rate at around 90%. Due to the declining birth rate, understaffing is expected to become even more severe shortly.

The Self-Defense Forces are dispatched approximately 800 times a year for disaster relief. That is an average of 2.2 times a day. During the Great East Japan Earthquake, approximately 100,000 people, or 40% of the 240,000 personnel, were dispatched for 50 days. There is concern that in the event of a large-scale disaster such as a Nankai Trough earthquake or an earthquake directly hitting the Tokyo metropolitan area, the country's national defense will be stretched thin due to emergency response operations.

Given these problems and limitations, why do even those who believe that a Ministry of Disaster Prevention and Reconstruction should be established not propose establishing a specialized disaster relief organization to replace the Self-Defense Forces? This is because disaster relief contributes to improving the public image of the Self-Defense Forces, gaining public acceptance, and recruiting more personnel.

However, the top priority should be to establish a system that can quickly and appropriately deploy the most necessary means to where they are most needed to save lives.

In an era in which disasters have become a part of daily life, Japan should establish a new Ministry of Disaster Prevention and Reconstruction and, under it, set up a specialized disaster relief organization to replace the Self-Defense Forces' disaster relief functions.

## 4 Critical Verification of the Reconstruction Processes and Outcomes

The fourth lesson underscores the need for a comprehensive and critical review of the reconstruction processes and outcomes. Fourteen years have passed since the Great East Japan Earthquake, and evaluating the recovery from the disaster remains a highly controversial topic. There are 'reviews' that boast of their achievements, but they lack the process of critical verification by a third party or experts. Therefore, they are what should be called one-sided results reports. For instance, *The Status and Efforts of Recovery from the Great East Japan Earthquake* (2024, with each year's edition) by the Reconstruction Agency, and *The Review of Local Recovery Development: Reconstruction from the Great East Japan Earthquake: Initiatives and effects of coastal disaster-stricken municipalities* 169 pages long by the Miyagi Prefecture Civil Engineering Department Urban Planning Division (2022) are typical examples.

On the other hand, Iwate Prefecture's *The Recovery from the Great East Japan Earthquake and Tsunami: Proposals from Iwate* (2020) is 379 pages long, and the content is also worth reading. It includes a message from Governor Takuya Tasso and an index at the end for ease of use. Neither the Reconstruction Agency nor the Miyagi Prefecture review has indexes and is more like a simple data collection. In Iwate Prefecture, the Reconstruction Promotion Division of the Reconstruction and Disaster Prevention Department controls the secretariat. Since the beginning of the disaster, the Iwate Prefecture Great East

Japan Earthquake and Tsunami Reconstruction Committee has been formed, comprising representatives from various fields in the prefecture, and the governor also attends every meeting. The committee meets approximately twice a year and has met 38 times by June 2025. Under the Tsunami Reconstruction Committee, a Comprehensive Planning Special Committee met totally 34 times, and a Women's Participation Promotion Special Committee met totally 22 times by the end of March 2023. All these meetings are held publicly, and meeting materials and minutes are also available on the prefecture's website. The editing of Iwate Prefecture's review was discussed at the 2019 meetings.

As part of the Iwate Prefectural Citizen Plan (2019–2028), the 1st stage reconstruction promotion plan (2019–2022) and the 2nd stage reconstruction promotion plan (2023–2026) will also review the progress management and verification of the reconstruction process. It continues regularly under these committees.

In contrast, Miyagi Prefecture held the Miyagi Earthquake Recovery Conference only four times between May and August 2011, with two of these conferences held in Tokyo. After 2012, no review of the verification with experts was conducted at all. Verification of reconstruction was only carried out by prefectural officials. The expert meeting at Miyagi Prefecture regarding the Great East Japan Earthquake has been minimal since 2012. The only meetings were the Miyagi Prefecture Earthquake Remains Expert Meeting (held seven times from 2013 to 2014) and the Expert Meeting on How to Pass on the Memory and Lessons of the Great East Japan Earthquake (held six times in 2017). No one can verify the Miyagi Prefecture Earthquake Recovery Plan announced in October 2011, which covered the period to fiscal 2020. The prefectural top officials do not recognize the necessity or understand the significance of reviewing this plan by experts.

Prefectural citizen awareness surveys have been conducted every year in both Miyagi and Iwate prefectures, but what is particularly noteworthy in Iwate prefecture is that as part of the progress management of the reconstruction promotion plan, surveys have been conducted by 152 participants (in principle, the same people each time) who live or work in 12 coastal municipalities. The "Iwate Recovery Watcher Survey" is conducted every year with respondents being "people who are in a position to be able to observe recovery efforts in the disaster-stricken areas," such as community association/organization executives, school teachers, and financial institution personnel. These have been carried out two or four times per year. There is a clear time gap in the sense of recovery between the southern part of the coast (south of Miyako City), which was more severely damaged, and the less damaged northern part of the coast (north of Iwaizumi Town). This study clearly shows Iwate Prefecture's stance of asking local leaders to evaluate reconstruction progress.

Why is critical verification by a third party or experts necessary?

Ultimately, the verification report emphasizes the importance of budget accountability. It is essential to ensure that the allocated funds are used effectively and transparently, instilling a sense of financial responsibility and trust in the disaster management and reconstruction processes.

There are differences of opinion and conflicts of interest regarding earthquake recovery planning, such as the construction of new seawalls, town revitalization, collective relocation promotion projects, the revitalization of industries and livelihoods, and the consolidation or closure of schools and hospitals, among others. Lessons learned and issues should be highlighted along with the results, and improvements should be made promptly.

Second, the process of reconstruction cannot be ended. As long as the area continues to exist, the success or failure of the reconstruction will continue to be questioned, regardless of how many years pass. Verification is indispensable and essential. If new local development and revitalization issues arise, it will be necessary to review and adjust the reconstruction trajectory accordingly. Criticism is also necessary for follow-up and ongoing review of the reconstruction.

The third reason why critical verification by a third party is important is to ensure credibility. The success or failure of the earthquake recovery process depends on the trust of the sufferer. Third-party verification can enable information to be made public and contribute to securing the trust of interested parties and local communities.

Fourth, understanding what went well and what did not in the earthquake recovery process will help improve future disaster preparedness and recovery plans, allowing us to learn valuable lessons. In preparation for the next major disaster, such as the Nankai Trough or an earthquake directly hitting Tokyo, we must humbly and sincerely learn the lessons. Therefore, we can refer to preparation for reconstruction before the next major disaster as "advance reconstruction." Today, if we anticipate disasters in advance, we can minimize damage and expedite the reconstruction process. There is a growing need for "advanced reconstruction" in the event of a huge disaster. To develop a comprehensive plan for a disaster-resilient region and foster consensus, it is crucial to summarize the lessons learned from the recovery efforts following the Great East Japan Earthquake. We should not miss it.

Along with the climate crisis, in recent years, flood damage, including landslides and flooding caused by localized torrential rains or the approach and landfall of large typhoons, has occurred almost every year in Japan. This trend has become even more pronounced over the past decade and a half. "Disasters occur frequently." Critical examination is essential in dealing with "frequent disasters."

Fifth, the estimated reconstruction-related budget execution from FY 2011 to FY 2021 is 39,448.2 billion yen (approximately $ 272 billion). When divided by the total population of 125.5 million people in Japan, the national expenditure per person for reconstruction is approximately 314,328 yen ($2,168). Many projects have been criticized for being oversized or overspending. The Reconstruction Agency and prefectural authorities are accountable to the public for the proper execution of their budgets.

## 5 Negative Effects of the Great Heisei Mergers

The fifth lesson is the adverse effects of "the great Heisei mergers" from 1999 to 2006. Keeping in mind the comparison with the 1995 Great Hanshin-Awaji Earthquake, let us point out some characteristics of the recovery process following the Great East Japan Earthquake.

I. Because the affected area was a depopulated and aging region, "creative recovery" was a symbol in the recovery from the Great East Japan Earthquake, and the aim would be to "achieve a recovery that will become a model for regional revitalization nationwide." Consolidation of residential areas, schools, and medical facilities, among other initiatives, also serves as a model for nationwide village reorganization projects.

II. "The great Heisei mergers" from 1999 to 2006 also had significant adverse effects. A branch office replaced the city office function after the merger. The impact of a 20% reduction in city employees is also severe. Areas that merged by 2006, such as Kesennuma City, Minamisanriku Town, Ishinomaki City, Higashimatsushima City, and Yamamoto Town in Miyagi Prefecture, were affected by the disaster while they were in the process of forming a new regional identity as a new city or town. Citizens of the peripheral, minor areas affected by the mergers have felt that they have suffered loss and dissatisfaction due to the mergers, such as difficulty in receiving disaster relief and support, delayed recovery, and inadequate media coverage. The great Heisei mergers were the result of neoliberal policies, which prioritized economic efficiency over community life and sustainability.

III. The areas affected by the Great East Japan Earthquake have suffered from a significant population outflow for many years. This outflow, influenced by various factors including the allure of the Tokyo metropolitan area and regional core cities like Sendai City, serves as a stark indicator of the extent of the damage caused by the earthquake and the urgency of disaster recovery.As shown in Table 3.1, when comparing the census

TABLE 3.1 The 2010 and 2020 census population of the disaster-affected municipalities in Miyagi Prefecture

| | 2010 | 2020 | Rate of change |
|---|---|---|---|
| Kesenuma City | 73489 | 61147 | -16.8% |
| Minamisanriku Town | 17429 | 12225 | -29.9% |
| Ishinomaki City | 160826 | 140151 | -12.9% |
| Onagawa Town | 10051 | 6430 | -36.0% |
| Higashimatsushima City | 42903 | 39098 | -8.9% |
| Matsushima Town | 15085 | 13323 | -11.7% |
| Rifu Town | 33994 | 35182 | 3.5% |
| Shiogama City | 56490 | 52203 | -7.6% |
| Tagajyo City | 63060 | 62827 | -0.4% |
| Tomiya City | 20416 | 18132 | -11.2% |
| Sendai City | 1045986 | 1096704 | 4.8% |
| Natori City | 73134 | 78718 | 7.6% |
| Iwanuma City | 44187 | 44068 | -0.3% |
| Watari Town | 34845 | 33087 | -5.0% |
| Yamamoto Town | 16704 | 12046 | -27.9% |
| Miyagi Prefecture | 2348165 | 2301996 | -2.0% |

SOURCE: CENSUS POPULATION

TABLE 3.2 The 2010 and 2020 census population of the disaster-affected municipalities in Iwate Prefecture

| | 2010 | 2020 | Rate of change |
|---|---|---|---|
| Hirono Town | 17913 | 15091 | -15.8% |
| Kuzi City | 36872 | 33043 | -10.4% |
| Noda Village | 4632 | 3936 | -15.0% |
| Fudai Village | 3088 | 2487 | -19.5% |
| Tanohata Village | 3843 | 3059 | -20.4% |
| Iwaizumi Town | 10804 | 8726 | -19.2% |
| Miyako City | 59430 | 50369 | -15.2% |
| Yamada Town | 18617 | 14320 | -23.1% |
| Otsuchi Town | 15276 | 11004 | -28.0% |
| Kamaishi City | 39574 | 32078 | -18.9% |

TABLE 3.2 The 2010 and 2020 census population of the disaster-affected (*cont.*)

| | 2010 | 2020 | Rate of change |
|---|---|---|---|
| Ofunato City | 40737 | 34728 | -14.8% |
| Rikuzen Takada City | 23300 | 18262 | -21.6% |
| Iwate Prefecture | 1330147 | 1210534 | -9.0% |

SOURCE: CENSUS POPULATION

TABLE 3.3 The 2010 and 2020 census population of areas in Ishinomaki City

| | 2010 | 2020 | Rate of change |
|---|---|---|---|
| Ishinomaki Area | 112683 | 99904 | -11.3% |
| Kahoku Area | 11578 | 9878 | -14.7% |
| Ogatsu Area | 3994 | 1031 | -74.2% |
| Kanan Area | 16950 | 18481 | 9.0% |
| Momoo Area | 7582 | 6629 | -12.6% |
| Kitakami Area | 3718 | 2135 | -42.6% |
| Oshika Area | 4321 | 2093 | -51.6% |
| Ishinomaki City | 160826 | 140151 | -12.9% |

SOURCE: CENSUS POPULATION

population of the municipality in 2010 before the earthquake and 2020 in the case of Miyagi Prefecture, there was a 36.0% decrease in Onagawa Town, a 29.9% decrease in Minamisanriku Town, and a 27.9% decrease in Yamamoto Town. The more damaged areas by the earthquake and tsunami, the more the population decreased. In areas relatively close to Sendai City, the population increased, and the decline was smaller.

Table 3.2 shows the affected municipalities in Iwate Prefecture, all of which have seen their population decrease by more than 10%. In particular, the population of Otsuchi Town has decreased by 28.0%, Yamada Town has decreased by 23.1%, Rikuzentakata City has decreased by 21.6%, and Tanohata Village has decreased by 20.0%.

In order to look in detail at what is hidden in population changes by the municipality, we take a closer look at Ishinomaki City, in which seven municipalities were merged into one city in 2005. Table 3.3 shows that in the Ogatsu district (formerly Ogatsu Town) on the coast, 74.2% of the population, or three-quarters, have left. In the Oshika district (formerly Oshika Town), 51.6% of the population has left; in the Kitakami district (former Kitakami Town), 42.6% has left. These are evidence of the extent of the damage in the area and the difficulty of reconstruction. Many residents gave up on living in their original area because it took time to form an agreement and secure land for relocation to higher ground. Especially in the Ogatsu area, 80% of the homes were destroyed or washed away. The high ratio of the decrease is a sign of the deep divisions and rifts within the area over the relocation to higher ground and the construction of seawalls.

Miyagi Prefecture's "The Verification of Reconstruction Town Development" seems to ignore the difficulties of recovery, the depth of splits and rifts within the area, and the weight of what the disaster victims have lost.

## 6 On-Site Reconstruction vs. Collective Relocation

On-site reconstruction or collective relocation and how to design "the disaster risk areas" also caused severe controversy. Which should have the priority to think about the recovery? Which should be the primary principle for recovery. This issue is the sixth lesson to describe. Designing disaster risk areas is a prerequisite for land readjustment projects and collective relocation projects for disaster prevention. There are various ways to designate areas, but listening to residents' wishes and ensuring consistency with the municipalities' relocation projects is essential. According to the Hyogo Earthquake Memorial 21st Century Research Organization Research Institute's *Current Status and Issues in Land Use Regulations After Natural Disasters* (2015), interviews with affected municipalities. It is said that a vast area had to be designated in order to carry out land readjustment projects and collective relocation projects.

Of the approximately 47,100 hectares of flooded area in the three earthquake-affected prefectures of Iwate, Miyagi, and Fukushima, exactly one-third (33.4%), or 15,700 hectares, was designated as a "disaster risk area" under Article 39 of the Building Standards Act. The sufferers were prohibited from constructing new or expanding residential buildings and were forced to give up rebuilding their homes in their current areas. The area of 15,700 hectares is 2.4 times the area inside the Yamanote Line of the core area of Metropolitan Tokyo, which is 6,440 hectares. As shown in Table 3.4, in Ofunato City, the ratio of disaster risk areas to the flooded area was as high as 96.3%. In the northern part of Miyagi Prefecture, where the Ria coastline is widespread, the ratio is high at 75.0%. The rate is low at 29.9% in central and southern Miyagi Prefecture. Among them, Yamamoto Town has the highest rate at

81.0%. In Table 3.4, Rikuzentakata City has an exceptionally low disaster risk area designation rate of 5.3%. This is because other local governments have designated more expansive disaster risk areas, whereas Rikuzentakata City has specified minimal areas, such as near-slope projects.

In the case of the broader designation of disaster risk areas, it took time to secure higher ground and build consensus on the selection of relocation sites, which brought severe conflicts between those who relocated in groups and those who chose to rebuild independently elsewhere. There are many cases where relocation to higher ground became a hindrance, and the community could no longer maintain itself.

A researcher estimated that collective relocation projects typically take two and a half years to initiate and another 580 days to complete. This means it took more than four years to finish the relocation. Construction and housing land costs have skyrocketed, and the total project cost was approximately 40 million yen (approximately $276,000) per unit. In particular, in the Ogatsu, Oshika, and Kitakami districts of Ishinomaki City, the cost averages 60 million yen (approximately $413,800) per unit. In 324 districts across 27 municipalities nationwide, which were affected by the 2011 disaster, land preparation for housing was completed through collective relocation projects by the end of March 2020. The approximately 600 billion yen (approximately $ 4.14 billion) for the projects was spent entirely from national funds.

If disaster risk areas were designated in a more limited way, and residents were allowed to rebuild their homes in their original areas, even in flooded areas, it would be convenient for them to rebuild their homes in their original areas in a much shorter period, and it would be easier for the community to survive. Complex construction works based on designating disaster risk areas and relocating to higher ground is not the only card for "disaster prevention and mitigation." Disaster prevention and mitigation are essential considerations that should be integrated with daily evacuation preparation and planning.

One method attracting attention for making early, small-scale relocation possible is the "insertion-type prevention project" realized in Ofunato City. It is a method of relocating vacant land in an existing village. In Ofunato City, using this method, 186 homes, or approximately half of the residents, were relocated, and the project was completed for 32.37 million yen (approximately $223,241) per home. Because large-scale construction work was no longer necessary, the project was completed in around 200 days. Other large-scale development projects took time to complete, so young people who could rebuild on their own were impatient and left in search of new employment opportunities. In contrast, older people who cannot find new jobs often live in remote areas. There are pessimistic concerns about the long-term sustainability of the affected area.The number of homes planned to be relocated to higher ground in the three earthquake-affected prefectures of Iwate, Miyagi, and Fukushima was approximately 28,100 as of December 2012. However, as of September 2017, five years later, it was approximately 18,300 (35% decrease from 2012). Iwate Prefecture had about 7,500 houses (35% decrease from 2012), Miyagi Prefecture about 9,000 houses (42% decrease from 2012), and Fukushima Prefecture had about 1,900 houses (26% decrease from 2012), to Igarashi et al. (2021).

Generally speaking, the principle of the on-site reconstruction should have priority. This principle can save time for consensus building, the budget for recovery, and a sustainable community.

## 7 The Successive Transition of Structural Strain and Disaster Areas as the Downstream

Diagrammatically, the problem situation above is as follows.

> The wide-area designation as a disaster risk area → The large-scale disaster prevention project → The prolonged consensus building and difficulty in securing relocation sites → The delay in starting prevention work → The large-scale civil engineering work → The leaving of young people → The difficulties in rebuilding a community on the newly developed land → The pessimistic concerns about the sustainability of the suffered area

This diagram illustrates the vicious circle of difficulties that sufferers and disaster areas had faced. This is a typical case of the 'successive transition of structural strain (Funabashi, 1988, p. 180).' The mechanism which new problems emerge one after another is that "measures to solve one problem create a new problem, and the next measure to solve it creates another problem." Such successive transitions of structural strain often happen in the case of the downstream. I have proposed environmental sociology as a 'sociology of the downstream' (Hasegawa, 2000). I generalized and referred to the production and distribution processes as upstream and the processes after consumption as downstream. When things are burned, carbon dioxide is emitted. At normal concentrations, such as those found in the air, carbon dioxide is harmless. For a long time, we have not been concerned about the problem that the concentration of carbon dioxide in the atmosphere is increasing gradually. Similarly, they were indifferent to the problem of radioactive waste produced by nuclear power. In this light, modern society is faced with various kinds of 'cleaning up' problems, to put it that way. It was an illusion to think that if you put it out in the garbage or flush it down the toilet, 'it is over.' It just disappeared on the surface in front of our eyes. Like waste, environmental bads or bads that place a burden on the environment are the cause of many of the troublesome problems facing modern society. The process by which environmental bads are emitted and processed is called downstream. We can apply the concepts of upstream and downstream to all aspects of life, discussing pollution problems as 'downstream' issues and positioning waste incineration plants and sewage treatment plants as 'downstream facilities.' General and industrial wastes, radioactive wastes, noise, and vibration, as well as dioxin, environmental hormones, and greenhouse gases, are negative environmental assets that cause environmental degradation. When development activities destroy historical landscapes and townscapes, as well as natural beaches, the development activities themselves should also be considered environmental bads. Thus, environmental bads are all materials, energy, and actions that place a burden on the environment. Conventional sociology, as well as conventional economics, has primarily focused on upstream issues, such as production and living processes. Although the management of environmental risks appears to have been refined on the surface, the downstream periphery, where the costs are concentrated, continues to be plagued by site difficulties, a lack of disposal sites, conflicts, and other problems. With globalization, the periphery of developing countries has been targeted as the end of the downstream, and environmental liabilities are imposed on the periphery. Modern society has been sustained by massive emissions of industrial and radioactive waste, as well as carbon dioxide, and now it is poised to be undermined by the environmental risks posed by the increasing concentration of carbon dioxide in the atmosphere and waste that has nowhere else to go.

In the context of disaster study, we can regard disaster areas located in the periphery, characterized by poor transportation services and limited shopping facilities, as the downstream of society. Conversely, the central area, where wealth is concentrated, has convenient transportation, and normal areas are considered upstream. It is the main reason disaster areas face the vicious circle described above and many serious problems.

## 8 Major General Contractor Smiles at Large-Scale Public Works Project

Who will benefit most from reconstruction through public works projects and infrastructure development? This is the seventh lesson.

Figure 3.1 shows the trends in the stock prices of Kajima and Taisei Corporation, two of the largest general contractors in Japan. During the

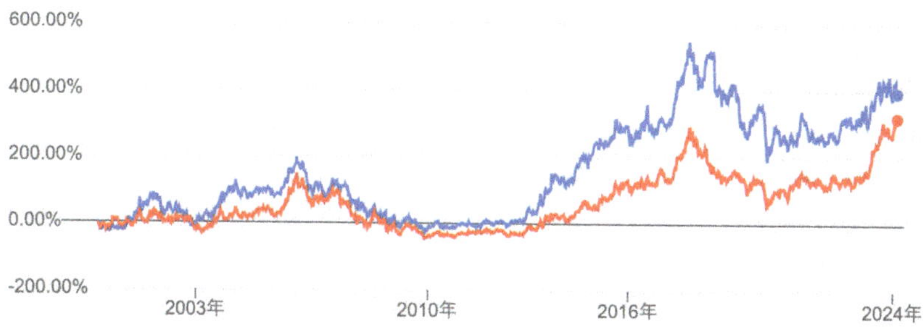


FIGURE 3.1 Trends of the stock prices of the largest general contractors
Red: Kajima Corporation
Blue: Taisei Corporation
SOURCE: NIKKEI INC.

Jun'ichiro Koizumi cabinet from April 2001 to September 2006, the two companies drastically reduced public works orders. The stock prices of both companies had been sluggish. The stock prices of both companies started to rise in the wake of the Great East Japan Earthquake, and they reached a particularly marked upward trend in December 2013 with the enactment of the Basic Law for Building National Resilience. Around the end of 2019, as the 10th anniversary of reconstruction approaches, the stock prices of both companies bottomed out. The trend in stock prices indicates that the Great East Japan Earthquake benefited major general contractors more than any other factor.

What kind of law is the National Resilience Basic Act? As a nod to the Great East Japan Earthquake, it is preparing for a Nankai Trough earthquake or an earthquake directly under the Tokyo metropolitan area. It is preparing for accidents to aging infrastructure, such as tunnel collapses, and improving railways and expressways. The Basic Act on National Resilience was enacted to secure an alternative route, and a vast project costing 200 trillion yen is expected over the next ten years. The project will continue through the implementation of the consumption tax increase and the issuance of construction bonds. This law aims to maintain and expand these public works projects.

The National Resilience Promotion Headquarters, headed by the Prime Minister and comprising all Cabinet members, has been established. Approximately every five years, the Basic Plan for National Resilience is formulated as a guideline. It is used to rebuild aging roads, bridges, tunnels, and embankments, among other infrastructure. The government plans to conduct systematic inspections and "vulnerability assessments" to identify issues and weaknesses in disaster prevention measures, prioritize them, and advance resilience policies. The First Basic Plan for National Resilience was created in 2014, and the

Second Basic Plan was created in 2018 and was revised with the aim of cabinet approval in July/August 2023. Since then, an annual plan has been drawn up that summarizes the measures to be taken that year.

The author believes that the most significant lesson from the Great East Japan Earthquake lies in reviewing the reconstruction approach centered on large-scale public works projects such as land readjustment projects, flood protection projects, and sea walls, and making new principles for a sustainable future. However, as a result, the continuation and expansion of large-scale public works projects was justified by enacting the Basic Act on National Resilience in 2013.

## 9 Conclusion: New Problems Emerge One after Another in the Name of the Reconstruction, Called "Reconstruction Disaster"

On May 10, 2011, two months after the Great East Japan Earthquake, the government's "the Great East Japan Earthquake Reconstruction Planning Council" announced "the Seven Principles for Reconstruction Planning" under the slogan of "the creative recovery," not simple "restoration from the disaster." Principle 2 states that the basic principle is community-based reconstruction. The national government will support this by discussing the overall reconstruction policy and designing the reconstruction system. Principle 4 states that renewable energy will promote the construction of regions. However, Higashimatsushima City is the only disaster-stricken area in Miyagi Prefecture that promotes the utilization of renewable energy. All seven principles of the recovery seem to have been empty, especially the "community-based reconstruction" was empty.

Overall, the reconstruction process has been led by the national government and the prefecture, including the construction of giant sea walls and elevated roads, relocation to higher ground, group-relocation promotion projects, and land readjustment projects. It was a reconstruction dependent on large public works projects by major construction companies. It is questionable to what extent the diversity of the disaster-affected areas was taken into account. The scope for municipalities' initiatives was minimal. These principles did not include the words and phrases of affected citizens or those who suffered.

Principle 5 surprisingly emphasizes the need for simultaneous recovery from the great earthquake and revitalizing the Japanese economy. How do both link? It seems unclear why recovering the affected local area can promote the revitalization of the macroeconomy and vice versa. As it turned out, this principle led to the use of the reconstruction budget for purposes other than disaster relief in non-disaster-affected areas.

The slogan of "the creative recovery" has functioned as a magical phrase to justify the arbitrary use of the reconstruction budget.

As Yoshimitsu Shiozaki (2014) aptly termed it, "the reconstruction disaster," during the reconstruction process, new problems emerge one after another in the name of "the reconstruction," and the reconstruction itself can bring about new "disasters." It also means 'successive transition of structural strain' described above. The reconstruction disaster is the eighth and final lesson of this paper. For whom and for what purpose was the reconstruction? Who is in charge of evaluating the reconstruction? The verification of recovery must be carried out comprehensively and from multiple angles, including the effects, support for disaster victims, the recovery of homes and towns, the recovery of industry and livelihoods, and the relationship between central areas as upstream and peripheral areas as downstream. The ideals of reconstruction, the nature of decision-making and policy-making, the merits and demerits of the reconstruction system, points for improvement, and cost-effectiveness should be examined.

## Bibliography

Aldrich, Daniel P. 2019. *Black Wave: How Networks and Governance shaped Japan's 3/11 Disasters*. Chicago: University of Chicago Press.

Funabashi, Harutoshi. 1988. "Kozoteki Kincho no Rensateki Ten'i (Successive Transition of Structural Strains)." In Funabashi et al. 1988. *Kosoku Bunmei no Chiiki Mondai: Tohoku Shinkansen no Kensetsu Funso to Shakaiteki Eikyo* [Regional Problems in a High-Speed Civilization: The Dispute over Construction of the Tohoku Shinkansen and its Social Impacts]. Tokyo: Yuhikaku. 155–187.

Hasegawa, Koichi. [2000] 2004. "Sociology of Environmental Issues: A Look at the Downstream Side." In Hasegawa. 2004. *Constructing Civil Society in Japan: Voices of Environmental Movements*. Melbourne: Trans Pacific Press. 20–33.

Hasegawa, Koichi. 2014. "The Fukushima Nuclear Accident and Japan's Civil Society: Context, Reactions and Policy Impacts," *International Sociology*, 29–4: 283–301.

Hasegawa, Koichi. 2015. *Beyond Fukushima: Toward a Post-Nuclear Society*. Melbourne: Trans Pacific Press.

Hasegawa, Koichi. 2018. "Continuities and Discontinuities of Japan's Political Activism before and after the Fukushima Disaster," In D. Chiavacci and J. Obinger eds., *Social Movements and Political Activism in Contemporary Japan: Re-emerging from Invisibility*, Oxford: Routledge, 115–135.

Hyogo Earthquake Memorial 21st Century Research Institute. 2015. *Shizen Saigai go no Tochiriyo no Genjyo to Kadai* [Current Status and Issues in Land Use Regulations after Natural Disaster]. Kobe: Hyogo Earthquake Memorial 21st Century Research Institute. (https://www.hemri21.jp/contents/images/2019/06/shizensaigaigonoto chiriyou_5082_5082.pdf, January 31 2024).

Igarashi, Takayoshi et al. 2021. *Fukko Jyunen no Sotenken: Sozoteki Fukko ni mukete* [Overview of 10 Years of Earthquake Recovery: Toward a "Creative Recovery"]. Tokyo: Iwanami Shoten.

Iwate Prefecture. 2020. *Higashi Nihon Daishinsai Tsunami karano Hukko: Iwate karano Teigen* [Recovery from the Great East Japan Earthquake and Tsunami: Proposals from Iwate]. Morioka: Iwate Prefecture. (https://www.pref.iwate.jp/_res/projects/default_project/_page_/001/027/741/fukkou_teigen_i_all.pdf, January 31 2024).

Miyagi Prefecture Civil Engineering Department Urban Planning Division. 2022. *Hukko Machi Zukuri no Kensho: Higashi Nihon Daishinsai karano Hukko, Engan Shicho no Torikumi to Koka.* [The review of local recovery development: Reconstruction from the Great East Japan Earthquake: Initiatives and effects of coastal disaster-stricken municipalities]. Sendai: Miyagi Prefecture. (https://www.pref.miyagi.jp/documents/38088/00hyoshi.pdf, January 31 2024).

Reconstruction Agency. 2022. *Higashi Nihon Daishinsai karano Hukko no Jyokyo to Torikumi* [The status and recovery efforts from the Great East Japan Earthquake]. Tokyo: Reconstruction Agency. (https://www.reconstruction.go.jp/topics/main-cat1/sub-cat1-1/202312_pamphlet_fukko-jokyo-torikumi.pdf, January 31 2024).

Samuels, Richard. 2013. *3/11 Disaster and Change in Japan.* Ithaca, NY: Cornel University Press.

Shiozaki, Yoshimitsu. 2014. *Fukko* ⟨*Saigai*⟩ [The reconstruction ⟨disaster⟩ ]. Tokyo: Iwanami Shoten.

Tanaka, Shigeyoshi. 2023. "The 2011 Great East Japan Earthquake and Tsunami: the Highest Casualties and Largest Reconstruction Funds—Characteristics of Major Disasters and Future Challenges in Developed Countries." *Japanese Journal of Sociology* 32(1): 7–24. (https://doi.org/10.1111/ijjs.12147/31, last accessed January 31 2024).

CHAPTER 4

# *Homo Emotionalis*—the Implicit Anthropology of Disaster Sociology

*Benoit Giry*

On April 18, 1906, William James, at the time a visiting professor at Stanford, experienced the great earthquake that ravaged the city of San Francisco. It was five-thirty in the morning when he was thrown to the ground. All the objects sitting on the furniture fell down as the tremors cracked the walls of the small apartment he occupied on campus. A few weeks later, on June 7, 1906, in *Youth's Companion*, he describes his state of mind at the time of the event:

> In my case, sensation and emotion were so strong that little thought, and no reflection or volition, were possible in the short time consumed by the phenomenon. The emotion consisted wholly of glee and admiration; [...]. I felt no trace whatever of fear; it was pure delight and welcome. '*Go* it,' I almost cried aloud, 'and go it *stronger*!' (James, 1987: 1215–1216)

In the hours and days that followed, inquiring about the reaction of his wife and the local residents, James noticed the general nature of this psychological state: pure emotion rather than reflection; excitement in the moment, followed by a special calm. Nowhere on the Stanford campus or in the city of San Francisco, tormented by its 3,000 dead, did he find any trace of fear or anxiety. He concluded that this *equanimity* was anthropological in nature:

> I like to think that what I write is a normal and universal trait of human nature. [...] Physical pain whether suffered alone or in company, is always more or less unnerving and intolerable. But mental pathos and anguish, I fancy, are usually effects of distance. At the place of action, where all are concerned together, healthy animal insensibility and heartiness take their place. At San Francisco, the need will continue to be awful, and there will doubtless be a crop of nervous wrecks before the weeks and months are over, but meanwhile the commonest men, simply because they are men, will go on, singly and collectively, showing this admirable fortitude of temper. (James, 1987: 1222)

 | HTTPS://DOI.ORG/10.1163/9789004751804-006

In this interpretation of disaster as an anthropological experience that reveals human nature, emotions appear to play a significant role. This is what seems to have intrigued James, and it is also what would intrigue his successors: what emotions does disaster generate? When they arise, to what extent do they determine an individuals' behavior? Finally, what practical and social effects do they have?

By emotion, we mean a mental and sensory event experienced by an individual who recognizes it and assigns it a positive or negative valence (James, 1884). Thus, most individuals are capable of feeling excitement, fear, joy, anger, surprise, sadness, etc. These states are generally private but are likely to produce typical physiological manifestations (such as crying, trembling, laughing, etc.) that function as signs in social interactions.

Emotions differ from "mood," that ordinary affective background, in their intensity and limited duration. They also differ in that they are triggered by a *stimulus*. Emotions have an object: we are afraid of, angry at, or surprised by something or someone. In a 1999 article, emotions sociologist Jonathan H. Turner argues that "emotions are aroused by the degree of congruity or incongruity between (1) what is expected and (2) what is experienced in a situation; and when there is high incongruity, emotional arousal ensues." (Turner, 1999: 134). And disasters, in the sociological sense, are defined precisely as a disruption of the normal course of social life (Fritz, 1961; Tierney, 2019). At first glance, everything seems to indicate that disasters are powerful generators of emotions.

Perhaps for this reason, sociologists studying disasters felt very early on the need to describe the emotional state of the people they were observing. In what is considered the first thesis on the sociology of disasters, Samuel Prince devoted an entire chapter to the emotional states caused by the Halifax explosion (Prince, 1920). Pitirim Sorokin's *Man and Society in Calamity* (1942) opens with a study of the effects of war, famine, revolution, and epidemics on the affective and emotional lives of individuals. This focus on the emotions of individuals and groups continues to this day.

Despite this long history, contemporary commentators refer to a recent "affective" or "emotional" turn in the field of disaster sociology. In 2025, at its congress in Rabat, RC39 questioned the affective turn in disaster sociology. This shift is part of a broader movement in sociology (Ticineto Clough et al., 2007) that now places emotions at the center of attention.

In this chapter, we defend the opposite thesis: there is no *affective turn* in disaster sociology; the emotional component has always been a structuring, even distinctive, element of sociological analysis of the actions of actors in disaster situations. Sociologists seek to understand how individuals cope with the emotions generated by disasters. Understood in this sense, disaster sociology can be conceived as an endeavor to develop a theory of action that gives emotions their rightful place. The corpus of disaster sociology seems to be inhabited by this implicit anthropological figure: *homo emotionalis*.

How did this figure come about? In broad outline, the construction of *homo emotionalis* and its integration into disaster analysis took place in three stages. Initially, sociology contributed to constructing a theoretical figure of the social actor in a disaster situation as an actor stripped of reason and socialization, driven by emotions. Second, research strongly questioned the empirical existence of this figure, though it remained an implicit theoretical reference against which the actual behavior of individuals was evaluated. In the third phase, in which we find ourselves today, sociologists are attempting to give *homo emotionalis* a more real consistency, which would allow theoretical reflection and empirical survey results to be articulated while integrating emotions into the sociological analysis of individual and collective action.

## 1 The Emergence of *Homo Emotionalis*

### 1.1 *The Fluidification of Social Institutions*

The sociology of disaster was born in 1917, in the aftermath of an explosion in the port of Halifax, Nova Scotia. On this occasion, the town's pastor, Samuel Prince, observed what he named a "sudden termination of the normal": the *City of Order* suffered 2,000 deaths, the destruction of its buildings, and a "fluidification" of its social institutions. This "fluidification" refers to the collapse of usual social reference points, such as class distinctions and norms, generated by the disaster:

> Old traditional social lines were hopelessly mixed and confused. The catastrophe smashed through strong walls like cobwebs, but it also smashed through fixed traditions, social divisions and old standards, making a rent which would not easily repair. Rich and poor, debutante and chambermaid, official and bellboy met for the first time as victims of a common calamity. (Prince, 1920: 32)

"Life becomes like molten metal," Prince wrote in the doctoral thesis he based on the event. "It enters a state of flux [...]. It is shaken perhaps violently out of rut and routine. Old customs crumble, and instability rules." (Prince, 1920:

18–19). This fluidification rendered the Halifax explosion a laboratory, a "strategic research site," in the words of Robert Merton, providing access to an anthropological reality:

> It is improbable that any single community has ever presented so composite a picture of human traits in such bold relief as appeared in the City of Halifax upon the day of the explosion. Human phenomena which many knew of only as hidden away in books, stood out so clearly that he who ran might read. (Prince, 1920: 35)

What are these human phenomena revealed by disaster? In other words, what remains of humans when the social institutions that organized their lives become fluid? For Prince, the answer is clear: disaster reveals human beings driven by their emotions.

Prince borrows the idea of emotional determination of action in disaster situations from the war neurology of Gustave Roussy and Jean Lhermitte and the crowd psychology of Gustave Le Bon. In the early 20th century, Roussy and Lhermitte worked on a particular type of post-traumatic stress suffered by soldiers in the World War I, known as "shell shock". This condition, which affected victims of bombing, manifested itself in episodes of a vacant smile, bulging eyes, mutism and/or deafness, and prostration. Contrary to the concussion theory, which claimed that the symptoms of shell shock were due to anatomical changes caused by the blast, Roussy and Lhermitte defended the "emotional shock" theory: the spectacle of destruction and death that is war causes an excess of emotions that produces the same effects as a physical shock. Roussy and Lhermitte write: "The soldier, profoundly impressed by the conditions and often experiencing the most distressing visceral disturbances from the emotion, reacts, not always voluntarily, but driven by that instinct deeply rooted in us all—that of self-preservation." (Roussy & Lhermitte, 1918: xxviii). Therefore:

> Much more than shock pure and simple, emotion is itself able to produce a neurosis or to localize and fix on some part of the body the psychological disorder, which is the very basis of the neuropathic phenomenon. The reply may be made that frequently concussion and emotion are both at work and that it is almost impossible to separate these two factors. We cannot agree with this. Concussion pure and simple, unaccompanied by any emotional element, rarely gives rise to neuropathic symptoms. (Roussy & Lhermitte, 1918: 156–157)

This effect is combined with the mechanisms of collective psychology theorized by Le Bon (1896). In crowds, suggestibility is greater and emotions are more easily transmitted. These emotions are also more easily converted into action according to the theory of Prince's thesis supervisor, sociologist Franklin H. Giddings: "Crowds are subject to swift contagion of feeling, they are sensitive to suggestion [...] and always manifest a tendency to carry suggested ideas immediately into action." (Giddings, 1916: 136). The emotional phenomena generated by the disaster then take on a magnitude that goes beyond the victims alone: the crowd acts as a sounding board, amplifying emotions and accentuating their behavioral effects.

Ultimately, it is these emotions that explain the behavior of social actors. During the catastrophe, "the human will [...] is in close association with the emotions." Emotion not only explains the type of action an individual will perform; it is also, more fundamentally, a condition for action. It is the emotion that compensates for the harmful physical effects caused by the disaster, such as fatigue, thanks to what Prince calls their "dynamogenic value." In disaster situations, "strong emotions may become an 'arsenal of augmented strength'" enabling disaster victims to overcome fatigue and tap into unusual resources to accomplish tasks they would not have thought themselves capable of (Prince, 1920: 52).

Therefore, the fluidification of social institutions caused by the disaster lays bare human nature. And when humans shed their social trappings, they reveal their "true selves," manipulated by their emotions. Thus, the thesis of the fluidification of social institutions gives rise to an anthropological figure: *homo emotionalis*.

### 1.2 *An Emotional Economy*

In the material and social context created by the disaster—this universe populated by *homo emotionalis*—the collective dynamics of societies depends on what we might call an "emotional economy." Indeed, very different emotions are likely to arise during a disaster: anxiety, fear, anger, joy, relief, gratitude, etc. These emotions can be classified into two groups: some have a destructive effect on social life; others, on the contrary, enable its recovery. "The stimulus of the same catastrophe, it thus appears, may result in two different types of responses—that of greed on the one hand or altruistic emotion on the other" (Prince, 1920: 51).

This observation led Pitirim Sorokin to develop the "law of diversification and polarization of the effects of calamity upon society." Sorokin first observed that disasters produce varying emotional effects depending on the biological, psychological, and social characteristics of individuals: "Some become brutalized, others intensely socialized. Some disintegrate—morally, mentally, and biologically; others are steeled into an unbreakable unity. In adversity some lose their sense of honor; others are ethically and spiritually reinforced" (Sorokin, 1942: 159).

This diversity does not stem from a random process of independent emotional events. According to Sorokin, every emotional reaction to a disaster triggers a counter-reaction. Consequently, while becoming increasingly diverse and radical, emotional reactions maintain "social equilibrium":

> 'Reactions' produced by [disaster] in one part of society are always counterbalanced to a certain extent by the 'counterreactions' in its other parts. Its destructiveness is paralleled by its creative stimulation. This generation of diverse and opposite reactions permits a given society and culture to undergo a calamity without being utterly destroyed, to preserve in part their integrity and balance, and to revive quickly after the calamity is ended. [...] Otherwise,—if the concrete effects of the disaster were all identical, and tended to be entirely negative, many a society would have been irretrievably destroyed by a single major catastrophe. (Sorokin, 1942: 159–160)

Therefore, Prince's propositions, like Sorokin's more systematic ones, suggest that the collective destinies of communities facing disasters depend on this emotional economy. An economy in which good, prosocial emotions must exist in at least as large a number as bad, antisocial ones in order to preserve the community's existence. This is the game in which *homo emotionalis* is involved, the regime to which social groups are subjected in disaster situations.

### 1.3 *The "Disaster Syndrome" and the Emotional Mechanisms of Catastrophe*

The idea that these emotions are organized into a non-random temporal, spatial, and social pattern appears relatively early in the corpus. As early as 1956, in his study of the Worcester tornado, anthropologist Anthony F.C. Wallace described what he called "the disaster syndrome":

> The disaster syndrome is a psychologically determined defensive reaction pattern. During the first stage, the person displaying it appears to the observer to be 'dazed,' 'stunned,' 'apathetic,' 'passive,' 'immobile,' or 'aimlessly puttering around.' [...] The second stage is one of extreme suggestibility, altruism, gratitude for help, and anxiousness to perceive that known persons and places have been preserved; personal loss is minimized, concern is for the welfare of family and community. [...] In the third stage, there is a mildly euphoric identification with the damaged community, and enthusiastic participation in repair and rehabilitation enterprises [...]. In the final stage, the euphoria wears off, and 'normally'

> ambivalent attitudes return, with the expression of criticism and complaints, and awareness of the annoyance of the long-term effects of the disaster. (Wallace, 1956: 109–110)

The fluidification of institutions resembles an uncontrolled form of Durkheimian collective effervescence: it plunges the actor into a temporary state of social levitation. Drawing on Victor Turner's work on ritual, disaster sociologists identify this transition phase as a state of "liminality": a transformative experience during which humans, temporarily stripped of their social statuses, reveal themselves in an irreducible anthropological form. The "disaster syndrome" describes this transition in four emotional phases, which appear to be generated by four distinct mechanisms: shock, compassion, empathy, and trauma.

The first and most controversial mechanism is shock. Shock refers to the immediate emotional reaction generated by the violence of the disaster. It is characterized by passivity and unusual calm among the victims. This calm is an emotional state in itself and not an absence of emotion. It is a direct effect of the disaster, leaving individuals stunned by the "horror" imposed by the spectacle of death or destruction (Hoffman, 2020: 157–158).

The second mechanism consists of a form of emotional communion that we could call "compassion." Charles Fritz argues that disaster allows for the public expression of emotions that are usually suppressed by prohibition or modesty; this expression contributes to their collective management, participating in the formation of a therapeutic community, a "*communitas*":

> The widespread sharing of danger, loss, and deprivation produces an intimate, primarily group solidarity among the survivors which overcomes social isolation, provides a channel for intimate communication and expression, and provides a major source of physical and emotional support and reassurance. The capacity of human societies under severe stress to contract from a highly elaborated set of secondary group organizations to a kind of universal primary group existence is probably their central built-in protective mechanism. (Fritz, 1996: 63)

This state only affects the immediate victims of the disaster, which is why we can talk about "compassion": individuals suffer together (etymologically, compassion comes from the Latin *compassio*, "to suffer with"). As indicated by one of Anthony Oliver-Smith's interviewees, a survivor of the 1970 earthquake in Peru, this shared suffering has the effect, of strengthening social bonds: "How we all suffered! How we all suffered together! How beautiful that life. How beautiful that life of suffering! Of brotherhood! We suffered together; even though it was terrible, it was also beautiful" (Oliver-Smith, 1992: 260).

The third mechanism is empathy: individuals well beyond the initial circle of the primary group mobilize to participate in the recovery of the affected community. It is no longer the spectacle of shared suffering but the suffering of others that generates empathy. This empathy is not an emotion but the process by which actors transmit an emotion to each other regardless of their actual practical experience. This emotional mechanism can generate large solidarity networks, as demonstrated by the channels for financing and implementing humanitarian actions.

Finally, the fourth phase is a return to a new normal. This new normal incorporates new constraints inherited from the consequences of the disaster. The long-term "annoyance" mentioned by Wallace may consist of material and physical difficulties, but they may also consist of "trauma," a psychological imprint left by the disaster. This trauma can go as far as the outright disappearance of the affected community (Erikson, 1976).

In this theory of "*disaster syndrome*" lies the greatest hope for empirical validation of the existence of *homo emotionalis*. Admittedly, disaster syndrome does not appear systematically in all actors. But, according to proponents of this theory, this is simply because their level of emotional sensitivity is likely to vary. However, everyone has their limits: "Some researchers argue that the qualities of the 'stressee' matter little—that is, 'every person has a breaking point.' What is critical then is the nature of the stressor" (Drabek, 1986: 148). Thus, whatever barriers our reason or socialization may establish between us and our emotions, there is a level of violence in the world that reveals the *homo emotionalis* within us.

## 2 Sociologizing *Homo Emotionalis*

### 2.1 *Self-Control as a Mechanism for Preventing the Emergence of* Homo Emotionalis

Starting in the 1950s, disaster sociology attempted to verify this anthropological hypothesis by systematically documenting the emotional state of social actors in disaster situations. What emotions do they feel? How are these emotions distributed among individuals? To what extent do these emotions guide their behavior?

This inventory of emotions experienced during disasters was first undertaken in 1954. That year, the National Opinion Research Center (NORC) at the

University of Chicago studied a series of tornadoes that ravaged White County, Arkansas. A group of 26 sociologists, trained for the occasion, was deployed to the disaster site and collected testimonies from survivors (342) and rescue workers (81). The collection of facts gathered on this occasion was of significance beyond this report: the NORC research would form the empirical matrix from which early work would draw to construct a sociological theory of disasters.

This early empirical work in disaster sociology was funded by the US military to study a strange illness. In 1948, in the town of Donora, air pollution generated by a local factory caused the immediate death of 20 people and made 7,000 others ill. Among those who fell ill, some reported symptoms without having been in direct contact with the polluting agent (Quarantelli, 1988: 284–285). This condition was rapidly conceived as an emotional contagion, an unwanted effect of empathy. The military feared that this process could have harmful consequences for the population in the event of an attack: if even individuals not directly affected by a destructive force are likely to be affected, then bombings could produce effects that are disproportionate to their actual destructive force, undermine the morale of the population, hinder the deployment of assistance to victims, and ultimately slow down the resilience process of the affected communities. The military was worried that people would get overwhelmed by their emotions. This emotional harm has a generic name: *panic*.

NORC researchers have identified several emotional states among disaster victims in White County: anxiety, fear, grief, and relief appear to be affecting social actors in disaster situations. Disasters are undeniably an emotional process. In line with theory, these emotions seem to follow a non-random temporal, spatial, and social pattern. Not all emotions appear in all individuals, in all places, or during all phases of the disaster. Nor does the same emotion produce the same effects in all individuals: gender and level of skills seem to play a role in the ability of actors to control their behavior (NORC, 1954; Barton, 1969). However, contrary to what the anthropological figure of *homo emotionalis* suggests, no individual appears to be overwhelmed by their emotions:

> The most outstanding aspect when the affective reactions of respondents during pre-impact and impact are considered, is the great degree of self-control that was reported. Nearly one-fifth of all the impact respondents stated that they were calm and unexcited. A considerable number of individuals did report they were either highly or mildly agitated, but only a very few stated that they lost self-control. (NORC, 1954: 103)

Admittedly, this disappearance of *homo emotionalis* in its anthropologically pure form may be linked to the survey methodology. As Bates *et al.* point out, "one of the major drawbacks in the studies [...] has been the reliance placed on the statements of the people themselves" (Bates *et al.*, 1963: 90). Consequently:

> the low proportion of individuals who stated that they were in a state of 'shock' or were stunned or dazed, may be underestimated since individuals' awareness of such conditions is probably less vivid than their awareness of more active affective states. [...] It may be, of course, that the very low proportion of individuals reporting loss of self-control, stems from a reluctance to report behavior deviating from the socially approved. (NORC, 1954: 103)

Therefore, researchers are aware of these limitations. However, "it should be noted that what respondents reported as the affective reactions of themselves fairly well parallels what respondents observed of the affective reactions of others. [...] Certainly, taking all the available direct and indirect evidence, there is little reason to doubt that respondents on the whole attempted to report their affective reactions truthfully." Thus, "generally speaking, the impression that emerges from the evidence is one of frightened persons engaging in relatively self-controlled behavior." (NORC, 1954: 103–104). The significant body of empirical work carried out in the 1960s, 1970s, and 1980s, fueled by the institutionalization and internationalization of sociological research on disasters, corroborated this hypothesis:

> There is some limited evidence of what has been called a 'disaster syndrome' (i.e., a state of shock leading to regression in normal cognitive processes). However, it appears only in the more traumatic and sudden kinds of catastrophes, is confined to the post-impact period, lasts only a short time (minutes or hours), and does not occur on a large scale. (Dynes & Quarantelli, 1973: 4)

The general principle that disasters generate strong emotions is not disputed. In disaster situations, individuals feel fear, anxiety, despair, anger, relief, renewed hope, and disappointment. However, it would appear that 1) the types of emotions generated by disasters vary depending on the context; 2) the "disaster syndrome," when it exists, is not so widespread among victims; 3) the emotions generated by disasters are not capable of destroying pre-existing social institutions or the volition.

These three caveats are important amendments to the concept of *homo emotionalis*. They enable us to break free from the trap of psychologizing social phenomena, into which the *reductio ad emotionis* has gradually led the sociology of disasters and the mechanical nature of the effects of disasters. Disasters are social phenomena: they do not have psychological effects without the mediation of social institutions and the rationality of actors.

In his inventory of sociological findings on disasters published in 1986, Thomas Drabek asks the following question: "How do disaster victims respond emotionally?" The answer is simple: most individuals respond in an appropriate and responsible manner; panic and uncontrollable agitation are rare. "There is a clear consensus on the question," he adds (Drabek, 1986: 146). This consensus—real, albeit relative—still holds today: panic and dysfunctional behavior indicative of emotional overload are relatively undocumented. In disaster situations, emotions run high but do not destroy social institutions or reason. The question, then, is less about how emotions manipulate social actors than about how social actors deal with their emotions.

## 2.2 *Emotions, Interests, and Norms*

The fact that emotions do not control human behavior in disaster situations does not mean that they do not play a decisive role in many social processes in disaster situations. Rather, they interact with the interests and norms that structure social situations. For example, it is not only the pursuit of material gain that drives some actors to loot in disaster situations, but the pursuit of material gain combined with low emotional investment. This is the case with the looting of security forces: "The security forces guarding a disaster area, of course, usually have maximal opportunity to engage in looting and minimal emotional involvement in the situation—a fact which logically suggests that their potentiality for looting is much greater than that of the general population." (Fritz & Mathewson, 1957: 54). Conversely, as Anthony Oliver-Smith points out, a strong emotional investment can cancel out the incentive effect: "People affirm who they are, and where they belong in the world is expressed far more through value and emotion than pure material interest." (Oliver-Smith, 2020: 233).

A paradigmatic example of the interplay between norms, interests, and emotions in determining social behavior is the phenomenon of convergence—the movement of external actors toward disaster sites. In his work on this subject, Charles Fritz emphasizes that certain emotions play a decisive role for a significant proportion of convergers. The "anxious," for example, converge because of their concern for their loved ones; they come to reassure themselves, but they also come to support them. However, this behavior is not solely determined by the emotions of the actors; it is supported by norms that existed prior to the disaster: "in virtually every society, there is a general social expectation that persons who have undergone a frightening and depriving experience need the presence of intimates for emotional reassurance" (Fritz & Mathewson, 1957: 40). This is also demonstrated by James Kendra and Tricia Wachtendorf in their book *American Dunkirk*, which deals with the evacuation of Manhattan Island during the September 11, 2001 attacks: while the spontaneous mobilization of recreational boaters and maritime transport professionals was indeed the product of emerging norms, these were largely facilitated by the dispositions acquired through the practice of navigation, the skills it imparts, and the values it conveys (Kendra & Wachtendorf, 2016).

Even though the phenomena of norm fluidification do exist, research shows that norms that fall out of use are quickly replaced by others. Social nature abhors a normative vacuum. The Emergent Norm Theory (ENT) attempts to explain the dynamics by which provisional norms emerge and become established in times of disaster. Initially formulated by two disaster sociologists, Lewis Killian and Ralph Turner, ENT predicts that in situations of radical uncertainty, such as those imposed by a disaster, this emergence occurs in two stages. First, individuals consult with each other in a process that sociologists call "milling." After a while, one or more individuals propose to stabilize a definition of the situation and what should be done: this is "keynoting." This process involves emotions at different stages. Support for the keynote speaker, for example, is based in part on charismatic legitimacy, which mobilizes emotions. In this sense, emotions contribute to the production of emerging norms: they define and reinforce them.[1]

But emotions alone are not enough. Contrary to what the *disaster syndrome* suggests, the ability of disaster emotions to generate social effects (such as the punishment of real or imagined perpetrators, political unification, or polarization) depends on many factors. In *After Tragedy Strikes*, Thomas Beamish shows that, to do so, disasters must be constructed as public tragedies. He highlights the social conditions that enable this mechanism to occur, such as the fact that victims must be recognized as worthy victims. The "trauma script" does not therefore impose itself automatically: its conditions of felicity depend

1 It is worth noting that the intellectual genealogy of the ENT can be traced back to the work of Gustave Le Bon, through its reception by the first Chicago school and the *green bible* of Parks and Burgess (Parks & Burgess, 1921). The ENT was therefore partly inspired by the same theories that gave rise to the figure of *homo emotionalis*.

on social norms or interests which, as a result, condition the social effects of disaster emotions (Beamish, 2024).

A proper understanding of action in disaster situations therefore requires taking into account the norms, interests, and emotions of the actors involved. In *Governing Affect*, Roberto Barrios argues that true recovery depends less on the ability of reconstruction actors to restore a material environment in a short period of time than on the restoration of the individual and collective feelings that existed before the disaster. Social reality has an important emotional dimension: feeling good involves living in a familiar social and material environment (Barrios, 2017). In the same vein, Rebecca Elliott's *sociology of loss* provides a synthesis that places the emotions of actors within a more complex behavioral pattern that encompasses the entire disaster life cycle. Thus, it is emotional attachment to a place that helps us understand why individuals stubbornly choose to live in areas recognized as dangerous by public authorities, to the detriment of their material interests and health. It is also this emotional attachment that helps us understand the individual and collective commitment of disaster victims in different areas who have had to mourn the loss of their neighborhood, their community, or their way of life (Elliott, 2021).

So, what should we do with the figure of *homo emotionalis*? Should we relegate it to the museum of outdated ideas, or can we use it to advance sociological research? In what follows, we attempt to show that this figure can help highlight the effects of disasters on the emotional socialization of actors and contribute to the effort to integrate emotions into sociological models of action.

## 3 Using *Homo Emotionalis* to Produce Good Sociology

### 3.1 *The Role of Disasters in the Emotional Socialization of Actors*

From a sociological perspective, works of art and disasters have several significant connections. Art is an important activity for the emotional socialization of members of a society: it is a way of sublimating, learning, and circulating emotions between individuals. Very early on, disasters became objects of contemplation and subjects of art. In its earliest uses in ancient Greek, the term "καταστρεφειν" ("catastrophe") was used to describe the dénouement of a play. The evocation of catastrophe to convey emotions is therefore undoubtedly a very ancient practice and continues in a wide variety of entertainment forms (art and disaster tourism both capitalize on the emotions generated by catastrophe, albeit in different ways and with different objectives).

It is undoubtedly Romanticism that took the association of art and catastrophe the furthest. Romantics valued disaster not only for its ability to generate emotions but also for its ability to represent the way the world works. We find this idea in Turner's paintings and Byron's poems, where disaster becomes a cosmological principle: disaster and emotional upheaval are two sides of the same coin and are not exceptions to normality; they are the principle by which the world works. This symbolic association between emotional turmoil and disaster is explicit in Lars Von Trier's film, where the planet *Melancholia* represents both the depressive state that damages social bonds and the "disaster"—etymologically, the "bad astral body"—that destroys life on Earth.

This association implies an aesthetic relationship with disaster: the world is sublime and terrible; sublime because it is terrible. Thus, catastrophe merges with art, and disaster is the manifestation of Creation as a work of art: "Storms of every sort, torrents, earthquakes, cataclysms, 'convulsions of nature,' etc., however mysterious and lawless at first sight they may seem, are only harmonious notes in the song of creation, varied expressions of God's love." (Muir, 1901). It is this mindset that explains the behavior of a romantic like John Muir, strangely reminiscent of James during the San Francisco earthquake:

> In Yosemite Valley, one morning about two o'clock I was aroused by an earthquake; and though I had never before enjoyed a storm of this sort, the strange, wild thrilling motion and rumbling could not be mistaken, and I ran out of my cabin, near the Sentinel Rock, both glad and frightened, shouting, 'A noble earthquake!' feeling sure I was going to learn something. (Muir, 1901)

Beyond their artistic representations, actual disasters are likely to indirectly alter the patterns of collective emotions by transforming artistic productions. They inspire literary creation, influence film production, generate graphic movements, and introduce modulations in the musical register (Cerulo, 1984; Scanlon *et al.*, 2012) well beyond "genre" productions. Art then becomes a way of prolonging the emotions generated by the disaster.

Through a feedback effect, these artistic productions can transform the representations of social actors. According to Mike Davis, Hollywood film production has completely transformed the perception of Los Angeles. It is no longer possible to understand what residents say about their city without considering this "geography of fear" (Davis, 1998). "Emotional geography" is thus likely to be affected by real or imagined disasters that strike territories.

Beyond emotions, the culturalist argument suggests that the experience of disasters can play a role in shaping the fundamental moods of members of a society. Gregory Bankoff attributes the mixture of fatalism and determination that Filipinos value under the name *bahala na* to their regular experience of disasters (Bankoff, 2003). Prior to this, British Japanologist Ivan Morris contrasted Western optimism with Japanese pessimism:

> This underlying pessimism comes from a combination of Mahayana Buddhism [...] with a peculiar liability to earthquakes and other natural catastrophes. It is significantly revealed in the Japanese fascination with the extreme situation, which has produced a long literature of disaster [...]. Yet in the very impermanence and poignancy of the human condition the Japanese have discovered a positive quality. Their recognition of the special beauty inherent in evanescence, worldly misfortune, and the 'pathos of things' (*mono no aware*) in many ways replaces the blithe Western belief in the possibility of 'happiness.' (Morris, 1975: 39–40)

Disasters are thus capable of transforming what Raymond Williams called the "structure of feelings" of a society: the collective and individual capacity to feel. There is every reason to believe that disasters contribute to the emotional socialization of social actors: directly, by offering individuals the opportunity to renew their attachment to a community by actively engaging in mutual aid practices; or indirectly, in the transfigured forms seen in works of art. Thus, "during disasters, societies lose wealth and members, but gain collective representations" (Giry, 2023: 84). These collective representations contribute to the construction of a culture. This culture is then incorporated by social actors in the form of dispositions within their habitus.

### 3.2 *Integrating Emotions into a Theory of Action*

Sociology as a science is a product of modernity, and this modernity has long been understood, in a Weberian vein, as a tendency toward increased rational action in the explanation of social behavior. Even though Weber considered emotional action to be a possible type of social action, the movement toward the "disenchantment of the world" has led to a decline in emotions as motives for action. Protestant ethics—characterized by a decline in magical beliefs in the Christian corpus—is an expression of this. It is itself conceived as an effort to control emotions in favor of reason: "Puritan asceticism—like all 'rational' asceticism—works to enable people to control their 'constant motives,' especially those which it 'instills' in them, in relation to their 'affects'" (Weber, 1920: 292, my translation). Modernity, accompanied by the construction of the state and the civilization of morals, is based on this decline:

> The moderation of spontaneous emotions, the tempering of affects, the extension of mental space beyond the moment into the past and future,

> the habit of connecting events in terms of chains of cause and effect—all these are different aspects of the same transformation of conduct which necessarily takes place with the monopolization of physical violence, and the lengthening of the chains of social action and interdependence. It is a 'civilizing' change of behavior. (Elias, 1998, p. 55)

However, the modernization process has not succeeded in producing a world in which actors carry out their social activities *sine ira et studio*, "without hatred or passion," in the manner of a Weberian bureaucrat. Since Weber, sociology has sought to show that the emotional repression brought about by modernity is not total. The sociology of emotions highlights the norms and reasons that help us understand the emergence of emotions in specific situations (Hoshchild, 1979; Katz, 1999).

There is therefore undeniably a sociology of emotions that informs us about how emotions are socially produced. But, for historical and undoubtedly epistemological reasons, it has had more difficulty conceiving of emotion as a cause of social facts. As Jack M. Barbalet wrote in the late 1990s, "this idea, that emotion is a social cause, is more likely to be resisted by sociologists than the idea that it is a social effect" (Barbalet, 1998: 9).

According to a somewhat conventional belief, disaster sociology is a "laboratory" for the social sciences. But a laboratory for what? Disaster sociology could be a strategic research site for studying the role of emotions in individual and collective action. It could support efforts that have been underway for many years (Heise, 1979; Scheff, 1990). As Anthony Oliver-Smith points out:

> Disasters are clearly periods in which people experience a vast spectrum of intense emotions—anxiety, fear, terror, loss, grief, gratitude, anger, frustration, relief, and resignation—in all their shadings and intensities. These emotions color their experiences of preparation (if any), the event itself, and the aftermath of a disaster for indefinite periods of time. They further help to shape their responses in regularly patterned ways, including organizing to carry out important tasks and tackle often insurmountable problems in the midst of enormous obstacles. However, the problematical construction of emotion in the current sociological analyses of collective behavior, of which response to disaster is but one example, has led to a focus on a more instrumental approach. Anthropological theory can correct this flaw and lead to better understanding of the interplay between sentiment and rationality in societies dealing with disasters and other crises, and in social life in general. (Oliver-Smith, 2020: 232)

•••

For some, such as Nicolas Demertzis, "the sub-fields of memory studies, trauma studies, and disaster studies have appeared in social science and humanities in tandem with the sociology of emotions" (Demertzis, 2020: xii). Without going that far, this chapter argues that disaster sociology has long been based on an implicit anthropological conception of humans in disaster situations as beings influenced by their emotions: *homo emotionalis*. Disaster sociology therefore very early on developed, albeit in an informal way, a tacit theory of emotional action.

The first attempt was maximalist: emotions were, in the ultimate analysis, what motivated actors in disaster situations; not their interests, not their socialization: their emotions. But this proposition posed several problems: 1. it opened the way to a psychologization of the analysis of social phenomena in disaster situations; 2. it was based on several sociologically suspect assumptions; 3. attempts to substantiate it empirically have consistently failed. Sociologists quickly concluded that while many emotions do emerge in catastrophic processes, these emotions are not the sole cause of social actors' actions, an uncontrollable force that manipulates them like puppets, *perinde ac cadaver*.

In a more reasonable version, *homo emotionalis* is a component of social actors who base their actions on the norms, interests, and emotions specific to a situation. Understood in this sense, the integration of emotions into the sociological understanding of social action in general remains a desirable goal. Indeed, in the two models that broadly organize theories of action in sociology, the rationalist model and the dispositional model, emotions are superfluous data with no real causal capacity. In rationalist models of action, emotions are excluded or reduced to dichotomous forms such as the "satisfaction/dissatisfaction" pair. In dispositional models, emotions are likely to be triggered in the course of action but are simply by-products. However, social actors do feel things and regularly justify their actions based on their emotions, particularly in disaster situations. For this reason, the sociology of disasters offers an extremely useful body of knowledge to contribute to the affective turn currently underway in the social sciences.

## Acknowledgments

I would like to thank Lucille Hagège and Juliette Quénéa for proofreading an earlier version of this text. This work has benefited from a government grant managed by the Agence nationale de la recherche as part of the France 2030 program, reference ANR-22-EXIR-0007.

## Bibliography

Bankoff, G., 2003. *Cultures of Disaster. Society and Natural Hazard in the Philippines.* Routledge, New-York.

Barbalet, J.M., 1998. *Emotion, Social Theory, and Social Structure. A Macrosociological Approach.* Cambridge University Press, Cambridge.

Barrios, R.E., 2017. *Governing Affect. Neoliberalism and Disaster Reconstruction.*

Barton, A.H., 1969. *Communities in Disaster: A Sociological Analysis of Collective Stress Situations*, Doubleday & Co, New-York.

Bates, F.L., Fogelman, C.W., Parenton, V.J., Tracy, G.S., 1963. *The Social and Psychological Consequences of a Natural Disaster: A Longitudinal Study of Hurricane Audrey.* National Research Council, Washington D.C.

Beamish, T.D., 2024. *After Tragedy Strikes. Why Claims of Trauma and Loss Promote Public Outrage and Encourage Political Polarization.* University of California Press, Oakland.

Cerulo, K.A., 1984. "Social Disruption and Its Effects on Music: An Empirical Analysis". *Social Forces* 62 (4), 885–904.

Davis, M., 1998. *Ecology of Fear: Los Angeles and the Imagination of Disaster.* Metropolitan Books, New York.

Demertzis, N., 2020. *The Political Sociology of Emotions. Essays on Trauma and Ressentiment.* Routledge, London and New-York.

Drabek, T.E., 1986. *Human System Responses to Disaster: An Inventory of Sociological Findings.* Springer-Verlag, New-York.

Dynes, R.R., Quarantelli, E.L., 1973. "The family and community context of individual reactions to disaster". *Preliminary Paper* #10, University of Delaware.

Elias, N., 1998. *On Civilization, Power, and Knowledge. Selected Writings.* University of Chicago Press, Chicago.

Elias, N., 1998. *On Civilization, Power, and Knowledge.* The University of Chicago Press, Chicago and London.

Elliott, R., 2021. *Underwater. Loss, Flood Insurance, and the Moral Economy of Climate.* Columbia University Press, New-York.

Erikson, K.T., 1976. *Everything in Its Path. Destruction of Community in the Buffalo Creek Flood.* Simon & Schuster, New-York.

Fritz, C.E., 1961. "Disaster", in: Merton, R.K., Nisbet, R.A. (eds), *Contemporary Social Problems*, Harcourt, Brace & World, New-York, 651–694.

Fritz, C.E., 1996. "Disasters and Mental Health: Therapeutic Principles Drawn from Disaster Studies". *Historical and Comparative Disaster Series* #10, University of Delaware.

Fritz, C.E., Mathewson, J.H., 1957. "Convergence Behavior in Disaster. A Problem in Social Control". *Disaster Study Number 9*, National Academy of Sciences—National Research Council, Washington D.C.

Fritz, C.E., Mathewson, J.H., 1957. *Convergence Behavior in Disasters: A Problem in Social Control.* National Research Council, Washington D.C.

Giddings, Franklin H., 1896. *Principles of Sociology. An Analysis of the Phenomena of Association and of Social Organization.* MacMillan & Co, London.

Giry, B., 2023. *Sociologie des catastrophes.* La découverte, Paris.

Heise, D.R., 1979. *Understanding events Affect and the construction of social action.* Cambridge University Press, New York.

Hochschild, A. R., 1979. "Emotion Work, Feeling Rules, and Social Structure". *American Journal of Sociology,* 85 (3), 551–575.

Hoffman, S.M., 2020. "'The Worst of Times, the Best of Times': Toward a Model of Cultural Response to Disaster". In: Oliver-Smith, A., Hoffman, S.M. (eds.). *The Angry Earth. Disaster in Anthropological Perspective.* Routledge, London and New-York.

James, W., 1884. "What is an Emotion?" *Mind* 9 (34), 188–205.

James, W., 1987. "On Some Mental Effects of the Earthquake". *William James: Writings 1902–1910.* Library of America, New-York.

Katz, J., 1999. *How Emotions Work.* University of Chicago Press, Chicago.

Kendra, J.M., Wachtendorf, T., 2016. *American Dunkirk. The Waterborne Evacuation of Manhattan on 9/11.* Temple University Press, Philadelphia.

Le Bon, G., 1896. *The Crowd: A Study of the Popular Mind.* London, T. Fisher Unwin.

Morris, I.I., 1975. *The Nobility of Failure: Tragic Heroes in the History of Japan.* Holt, Rinehart and Winston, New-York.

Muir, J., 1901. *Our National Parks.* Houghton Mifflin, Boston.

NORC, 1954. *Human Reactions in Disaster Situations,* Report n° 52, University of Chicago.

Oliver-Smith, A., 1992. *The Martyred City: Death and Rebirth in the Andes.* University of New-Mexico Press.

Oliver-Smith, A., 2020. "The Brotherhood of Pain: Emotion and Social Organization in the Crisis of Disaster". In: Oliver-Smith, A., Hoffman, S.M. (eds.). *The Angry Earth. Disaster in Anthropological Perspective.* Routledge, London and New-York.

Park, R.E., Burgess, E.W., 1921. *Introduction to the Science of Sociology.* University of Chicago Press, Chicago.

Prince, S.H., 1920. *Catastrophe and Social Change. Based Upon a Sociological Study of the Halifax Disaster.* PhD, Columbia University.

Quarantelli, E.L., 1988. "The NORC Research on the Arkansas Tornado: A Fountainhead Study". *International Journal of Mass Emergencies and Disaster* 6 (3), 283–310.

Roussy, G., Lhermitte, J., 1918. *The Psychoneuroses of War.* London: University of London Press.

Scanlon, J., Johnston, N., Vandervalk, A., Sparling, H., 2012. "101 Years of Mine Disasters and 101 Years of Song: Truth or Myth in Nova Scotia Mining Songs?". *International Journal of Mass Emergencies & Disasters,* 30 (1), 34–60.

Scheff, T.J., 1990. "Socialization of Emotions: Pride and Shame as Causal Agents". In: Kemper, T.D. (ed.). *Research Agendas in the Sociology of Emotions*. State University of New-York Press, Albany.

Sorokin, P.A., 1942. *Man and Society in Calamity*. Routledge, New-York, 2017.

Ticineto Clough, P., Halley, J. (eds.), 2007. *The Affective Turn. Theorizing the Social*. Duke University Press, Durham and London.

Tierney, K.J., 2019. *Disasters. A Sociological Approach*. Polity Press, Cambridge.

Turner, J.H., 1999. "Toward a General Sociological Theory of Emotions." *Journal for the Theory of Social Behavior* 29 (2), 133–161.

Turner, R.H., 1967. "Types of Solidarity in the Reconstituting of Groups". *The Pacific Sociological Review* 10 (2), 60–68.

Wallace, A.F.C., 1956. *Human Behavior in Extreme Situations*. National Academy of Sciences, Washington D.C.

Weber, M., 1920. *Wirtschaft und Gesellschaft*. J.C.B. Mohr, Tübingen.

CHAPTER 5

# Four Horsemen and the Red Queen: Intersectionality of Risks and Institutional Resilience in Taiwan

*Thung-Hong Lin*

## 1 Introduction

The Global Risks Report 2025 by the World Economic Forum highlights armed conflicts and extreme weather events as the most pressing global risks in the near future, from a survey of nearly a thousand international experts. Taiwan is identified in the report as vulnerable to both threats (Global Risks Report, 2025). Beijing considers Taiwan a renegade province of China, despite Taiwan's de facto sovereignty over its territory and population. The Taiwan Strait, one of the busiest maritime and air traffic corridors in the world, exposes Taiwan to threats from pandemics and geopolitical tensions. Furthermore, Taiwan's location at the intersection of the Eurasian and Philippine Sea tectonic plates and in the Pacific Ring of Fire renders it susceptible to earthquakes. Moreover, the Northwest Pacific monsoon frequently generates typhoons, leading to floods and landslides. Taiwan is thus subject to the compounded risks from these natural and geopolitical vulnerabilities.

Most mass-casualty events in Taiwan have been epidemics, wars, earthquakes, or typhoons—often likened to the Four Horsemen of the Apocalypse. With the development of the field of disaster sociology, Taiwanese scholars have analysed the loss in human wellbeing caused by these catastrophic events. Disaster-induced losses in human wellbeing are commonly measured in health and economic terms. The scale of loss reflects **disaster risk**, which is generally understood as the interaction between **hazards**, **exposure**, and **vulnerability**. Postdisaster recovery efforts serve to mitigate loss and exemplify the concept of resilience (Intergovernmental Panel on Climate Change, 2025).

In my previous research on Taiwan's earthquakes, I introduced the concept of institutional resilience, which refers to how fast and how much institutions change to better prevent and respond to disasters. A political community with a greater capacity for policy adaptation and institutional reform following a disaster is better able to mitigate the loss in human wellbeing in future catastrophes. By examining Taiwan's historical encounters with disasters, we

HTTPS://DOI.ORG/10.1163/9789004751804-007

can examine Taiwan's ability to withstand and recover from crises and examine how institutional resilience has evolved (Lin, 2022).

My research has identified several key determinants of institutional resilience. First, political institutions influence disaster governance outcomes. In a comparative study of 150 countries, I demonstrated that disaster-related mortality is associated with the characteristics of political institutions, particularly state capacity and civil society engagement. State capacity refers to a modern sovereign state's administrative power over its population and encompasses defence, law enforcement, and taxation (Hendrix 2010; Skocpol, 1985; Tilly, 1992). Civil society is composed of voluntary associations, such as religious and lineage organisations, industry and labour unions, community groups, and various social networks—collectively referred to by some scholars as a source of social capital (Heinrich, 2010; Putnam, 1993). State–society relations, often analysed in social science in terms of the distribution of political power and resources, has rarely been analysed in disaster research (Özerdem & Jacoby, 2006).

In strong democratic states, civil society plays an active role in disaster prevention and recovery, giving rise to a competitive yet cooperative dynamic with the state—a phenomenon termed the 'Red Queen Effect' by Daron Acemoglu and James A. Robinson (2019). Democracies with stronger state capacity tend to have lower rates of disaster-related mortality, and democracies with weaker state capacity tend to have fewer disaster preparedness resources. Authoritarian regimes suppress freedom of speech and association, hindering the flow of information and the mobilisation of resources in civil society (Lin, 2015). Socioeconomic inequality influences disaster governance. Higher income inequality correlates with higher rates of disaster-related mortality, including pandemic-related mortality (Oronce et al., 2020).

Building on the concept of the Red Queen Effect, institutional resilience can be defined as the capacity of a political community to adaptively evolve disaster mitigation, emergency response, and recovery institutions at a pace and to an extent sufficient to reduce losses in human welfare. In democratic contexts, pressures from civil society development and electoral competition tend to foster stronger institutional resilience. Moreover, higher levels of state capacity enable greater investment in disaster-related institutional reforms, thereby enhancing the effectiveness of risk reduction strategies. This also helps to explain why some authoritarian regimes with relatively strong state institutions are nonetheless able to significantly reduce disaster-related losses, despite the absence of electoral accountability. Conversely, socioeconomic inequality may erode social capital, diminishing the incentives among the wealthy and powerful to invest in risk reduction measures that benefit poorer populations. In this way, inequality undermines institutional resilience by weakening trust, constraining social cooperation, and impeding inclusive disaster governance (Lin et al., 2022).

Taiwan's disaster history serves as a classic case study of institutional resilience. Given the high frequency of earthquakes and typhoons in the island, state capacity and civil society dynamics in Taiwan have continuously shaped the ability of institutions to successfully reform themselves after a disaster, contributing to low levels of human wellbeing loss from disasters. This resilience was clearly demonstrated during the COVID-19 pandemic. Taiwan had one of the lowest rates of COVID-19-related mortality largely because of its extensive experience with infectious diseases, extensive state infrastructure, robust public health institutions, and active civil society engagement in pandemic response (Lo et al., 2024).

To illustrate Taiwan's case, Section 2 introduces several mass-casualty events in Taiwan. Using the EM-DAT database from the Université Catholique de Louvain, I identified disasters with a death toll exceeding 10 (excluding deaths from causes such as suicide, homicide, chronic disease, and traffic accidents) (Centre for Research on the Epidemiology of Disasters, 2025). To extend the institutional resilience framework to armed conflicts, I also analyse events with more than 25 casualties, following international standards (UCDP, 2025). Next, I examine Taiwan's three major disasters in the democratic era: the 1999 Chi-Chi Earthquake, Typhoon Morakot (2009, and the COVID-19 pandemic, discussing state capacity, civil society engagement, and institutional resilience in response to these crises. Finally, I explore how postpandemic geopolitical shifts in the Indo-Pacific region have heightened the risk of war in Taiwan, posing new challenges to institutional resilience in Taiwan with respect to the management of compound risks.

## 2 Taiwan: a Volatile Island in History

In 1582, a Portuguese junk set sail from Macau to Nagasaki, Japan, carrying more than 300 passengers, including Portuguese Jesuits, Spanish missionaries, Han Chinese sailors, Japanese traders, and people from Africa and the Philippines. The ship encountered a typhoon in the northern Taiwan Strait and ran aground at the Tamsui River estuary. This event marked the first recorded Western encounter with 'Formosa', as Taiwan was later known. At that time, the island was inhabited by Austronesian-speaking Indigenous peoples who had lived there for thousands of years and Chinese and Japanese fishermen and pirates. The stranded sailors repaired their ship with the help of Indigenous peoples. Upon leaving, the sailors suggested that their empires establish military fortifications on the island (Cheng, 2021; Weng, 2006). In 1624, the Dutch East India Company (Vereenigde Oost-Indische Compagnie, VOC) conquered southern Formosa, establishing the first colonial administration on the island. In 1662, Koxinga (Zheng Chenggong)—a pirate warlord of mixed Chinese and Japanese heritage—expelled the Dutch and founded a short-lived kingdom as part of his efforts to restore the Ming Dynasty (Andrade, 2008). Viewing Taiwan as a potential security threat, the Manchurian (Qing) Empire launched an invasion and annexed the island in 1683. From that point on, Taiwan experienced frequent uprisings and became a strategic outpost for various competing empires in East Asia (Ka, 2021).

A perfect storm blew European imperialists onto Taiwan's shores, triggering successive foreign invasions. These invasions were not mere coincidences but rather a reflection of Taiwan's historical role as a geopolitical fault line—a site of power struggles among competing empires. By the 19th century, as China's imperial power declined, the Qing Dynasty opened Taiwan's ports to Western

TABLE 5.1 Top 15 mass-casualty events in Taiwan in 1895–2024, ranked by death toll

| Rank | Year | Event | Death toll |
|---|---|---|---|
| 1 | 1918–20 | 1918 Spanish flu pandemic | 48,866 |
| 2 | 1895–1917 | Plague and malaria (and the others following Japanese occupied) | > 30,600 |
| 3 | 1947 | February 28 massacre and smallpox/cholera (KMT vs. Taiwanese) | > 20,000 |
| 4[a] | 2020–2024 | COVID-19 | > 19,005 |
| 5 | 1895 | Japanese invasion of Taiwan | > 14,164 |
| 6 | 1944–45 | The Pacific War (US vs. Japan/Taiwanese) | > 5,582 |
| 7 | 1935 | Shinchiku-Taichū earthquake | 3,276 |
| 8* | 1999 | Chi-Chi (921) earthquake | 2,444 |
| 9 | 1906 | Meishan earthquake | 1,258 |
| 10 | 1959 | Tropical Depression 081 (87 floor) | 1,075 |
| 11* | 2009 | Typhoon Morakot (88 floor) | 699 |
| 12 | 1958 | Second Taiwan Strait Crisis (PRC vs. ROC) | > 618 |
| 13 | 1954 | First Taiwan Strait Crisis (PRC vs. ROC) | > 519 |
| 14 | 1963 | Typhoon Gloria | 363 |
| 15 | 1961 | Typhoon Pamela | 279 |

a Occurred within the last 30 years.

trade. Both the United States and France attempted to invade Taiwan. Japan successfully seized the island from China in 1895, transforming it into a modern colony. The Japanese colonial government introduced modern governance, population and land statistics, and disaster science, marking the beginning of Taiwan's institutional approach to disaster management (Yanaihara, 2022). A list of mass-casualty events in Taiwan from 1895 to 2024 is provided in Table 5.1. Each event is either an epidemic, war or government crackdown (often accompanied by an epidemic), earthquake, or extreme weather event (e.g., typhoon) (Lin, forthcoming).

Japanese imperialism in Taiwan faced not only resistance from the local population but was also accompanied by severe public health challenges. In the early years of military occupation, the leading causes of death among both Japanese and Taiwanese were not armed conflicts (which killed more than 14,164 people) but rather plague and malaria. More than 30,600 individuals died because of plague and malaria during that time (Fan, 1995, 1996). Natural disasters, including earthquakes, fires, and typhoons, were also major causes of death (Cheng et al., 1999). In response, the Japanese colonial government launched extensive scientific research on diseases and natural disasters to mitigate the health risks of the island's colonial rulers. Over time, these scientific advancements and governance strategies also filtered down to the local population, leading to more comprehensive state intervention in disaster prevention. Consequently, Taiwan developed an unusually precise system of disaster mortality statistics, reflecting the increasing depth of colonial governance (Government-General of Taiwan, 2025). During the 1930s, Japan's militarism became more aggressive. Japan expanded its invasions into China. By 1941, Taiwan became a military outpost for Japan's launch of the Pacific War. Towards the end of World War II, large-scale American air raids were being conducted against Taiwan. These air raids resulted in the deaths of 5,582 individuals, once again entangling the island in global great power conflict (Chang, 2015).

The international order following World War II and geopolitical shifts altered Taiwan's fate. In 1945, following Japan's surrender, the Allied forces placed Taiwan under the administrative control of Chiang Kai-shek's Republic of China (ROC) government. By 1947, economic collapse and epidemics triggered public protests against the new regime. In response, the Chinese Nationalist (KMT) government launched a brutal crackdown known as the February 28 Massacre. If we include the smallpox and cholera epidemics that spread from China, more than 20,000 lives were lost during this period (Chen, 1992; Lai, 2015). In 1949, after the KMT's defeat in the Chinese Civil War, the ROC retreated to Taiwan. The regime narrowly repelled an attempted Communist invasion during the Battle of Kinmen (1949 (Chen 2019). In June 1950, the outbreak of the Korean War transformed Taiwan into a key United States military outpost against communism. This temporarily safeguarded Taiwan from invasion by the newly founded People's Republic of China (PRC. Nevertheless, this also subjected the island to decades of authoritarian rule and continuous military threats (Khan, 2024).

During the Cold War, as the KMT loosened its authoritarian grip and Taiwan transitioned to an export-oriented economy (which it did with United States assistance), the nature of mass-casualty events shifted. Most events during this period were earthquakes or typhoons. Taiwan's deadliest flood occurred on August 7, 1959. That flood killed 1,075 individuals (Huang, 2011). Severe typhoons in 1961 and 1963 also caused hundreds of deaths. These extreme weather events forced the KMT government to establish disaster response mechanisms, such as involving the military in disaster relief and strengthening the firefighting department (Chen, 2022).

Since the 1980s, Taiwan has undergone a political transition from authoritarianism to democracy. As the government enshrined freedom of association, civil society organisations gained the ability to mobilise resources and personnel for disaster relief, significantly reshaping Taiwan's emergency response and disaster management systems. Over the past 30 years, three major disasters that have had profound effects on Taiwan's institutional development have occurred—each comprising multiple events: the 1999 Chi-Chi Earthquake (along with Typhoon Toraji in 2001), Typhoon Morakot in 2009, and two coronavirus pandemics (SARS in 2003 and COVID-19 in 2020). These events not only tested Taiwan's state capacity but also strengthened its disaster governance and resilience mechanisms, shaping the island's approach to crisis management in the new century.

## 3 The Impact of the Chi-Chi Earthquake

At 1:47 AM on September 21, 1999, the Chi-Chi Earthquake struck Taiwan, with its epicentre in Chi-Chi Township, Nantou County, approximately 12.5 km west of Sun Moon Lake at a depth of only 8 km. Measuring 7.3 on the Richter scale, the Chi-Chi Earthquake was the most severe earthquake in Taiwan since the 1935 Shinchiku–Taichū earthquake, which killed 3,276 people. The Chi-Chi Earthquake resulted in 2,415 confirmed deaths, 29 missing persons (totalling 2,444 casualties), 11,306 injuries, and the collapse or severe damage of more than 89,347 houses, with estimated financial losses exceeding €10 billion at the time. As Taiwan's second deadliest earthquake in recorded history, it had a profound impact on Taiwan's disaster prevention and response systems, especially given that it occurred during Taiwan's transition to democracy. My collaborators and I examined the earthquake in terms of hazards, exposure, vulnerability, and post-disaster resilience. Regarding hazards, I analysed seismological data, including ground motion intensity and surface rupture data and the effects of the Chelungpu Fault, which was responsible for the earthquake's destructive force. Geological factors explained approximately 75% of the variance in mortality rates, and social factors explained the remaining 25% (Lin et al., 2015).

Exposure refers to the structural resilience of buildings and the number of occupants within them. Casualties were concentrated near the Chelungpu Fault and in townships with a high population density. Casualties also occurred in areas distant from the epicentre, raising questions about other risk factors. In total, 89,347 houses collapsed or were heavily damaged. The fatality rate for single-unit houses averaged 0.14 deaths per collapsed house and for condominium buildings (four floors or more with at least eight households) averaged 4.8 deaths per collapsed building. The deadliest single-building collapse was the Sungshan Building in Taipei, far from the epicentre, which killed 87 people, suggesting that poor construction quality and corruption in the housing market posed significant disaster risks beyond geological factors.

Mortality rates were high in suburban satellite cities surrounding the Greater Taichung Metropolitan Area, which I termed the 'nexus township'. This vulnerability was mainly due to substandard construction quality in high-rise buildings. The two primary factors leading to collapses here were outdated building codes and corruption in the construction industry. Furthermore, low-income populations were often housed in the most structurally compromised buildings. We found that areas with lower average income, high levels of population outflow, a higher proportion of rental tenants, ageing demographics, and a greater number of female residents had higher earthquake mortality rates. These factors related to social vulnerability limited the evacuation capacity of at-risk populations and increased exposure to poorly maintained housing (Chung et al., 2022).

The process of reconstructing housing after the earthquake was a test of community resilience. Mortality rates were highest among residents of condos. I compiled historical reconstruction data from 142 collapsed condos, finding that only 94 were successfully rebuilt. On average, the reconstruction process took 1,950 days (more than 5 years), with the longest case taking 5,479 days (15 years), after which no further buildings were reconstructed. What determined whether a collapsed building was successfully rebuilt or remained in ruins? Given that government incentives, such as floor area bonuses and interest subsidies, were uniform, our study found that the key factors for successful reconstruction were strong disaster survivor organisations, support from external civil society groups, and political attention and advocacy. Conversely, unfair reconstruction projects and corruption of community leadership often led to legal disputes among disaster survivors, delaying or even derailing reconstruction efforts. Ultimately, we found that 'social capital'—the ability to mobilise collective action—was the primary determinant of successful postearthquake housing recovery (Chao & Lin, 2019).

Research on the aftermath of the Chi-Chi Earthquake further demonstrated that state capacity and civic engagement played crucial roles in institutional resilience—a key advantage of democratic systems. Before the earthquake, Taiwan lacked a formal disaster prevention law. In response to the crisis, then-President Lee Teng-hui issued emergency executive orders to coordinate rescue operations, reflecting the arbitrary powers of authoritarian-style crisis management. Recognising this legal gap, the Legislative Yuan swiftly enacted Taiwan's first Disaster Prevention and Protection Act, which established the National Science and Technology Center for Disaster Reduction. This act was designed to enhance scientific disaster studies, legislated the legal authority of the military to participate in disaster relief without requiring emergency executive orders, and integrated public donations into legally regulated post-disaster relief funds, ensuring financial transparency and preventing fraud. This legal framework not only improved collaboration between civil society and the state but was later adapted for Taiwan's pandemic response efforts, demonstrating the long-term effects of institutional reforms driven by experience with disasters (Lin & Chen, 2022).

The Taiwanese experience underscores the critical role of state capacity in mitigating earthquake risks. Following the Chi-Chi Earthquake, Taiwan strengthened its building and fire codes through the National Center for Research on Earthquake Engineering. Additionally, earthquake insurance became mandatory for all new buildings and real estate transactions, reinforcing seismic resilience in construction. More than half of Taiwan's buildings were constructed before 1999 under outdated earthquake codes and remain susceptible to seismic events. To address this structural vulnerability, the government retrofitted all public buildings, including government offices, hospitals, and schools, ensuring that all public buildings complied with the new seismic safety standards. The results of this effort soon became evident. Three major earthquakes have occurred in the past decade. These are the 2016 Southern Taiwan Earthquake (USGS MW 6.4, 116 deaths), the 2018 Hualien Earthquake (USGS MW 6.4, 17 deaths), and the 2024 Hualien Earthquake (USGS MW 7.4, 18 deaths). No civilian casualties resulted from damage to public buildings in any of these earthquakes—all recorded fatalities were linked to the collapse of private structures built before 1999. These outcomes demonstrate the effectiveness of Taiwan's long-term seismic risk reduction strategies, reinforcing the importance of institutional resilience in disaster governance (Chung et al., 2022).

Research on the Chi-Chi Earthquake has shown that postdisaster institutional reforms significantly enhanced Taiwan's state capacity, particularly in strengthening building codes and retrofitting public infrastructure. However, because governmental subsidies to affected households were distributed uniformly, the pace and success of reconstruction in the private housing sector depended more heavily on the self-organisation and inclusive governance of civil society. This disparity in postearthquake recovery outcomes across different communities highlights the Red Queen Effect at work—where institutional resilience was shaped not only by the strength of the state but also by the capacity of civil society and local community to mobilise, negotiate, and collaborate effectively.

## 4 Climate Change: the Impact of Typhoon Morakot

In 2009, Typhoon Morakot made landfall in Taiwan, bringing record-breaking rainfall to central and southern regions and triggering severe floods and widespread devastation. The disaster resulted in 673 deaths and 26 missing persons (a total of 699 casualties) and agricultural losses exceeding €600 million. The disaster was Taiwan's most devastating typhoon in 50 years. The most catastrophic incident occurred on August 8 at midnight, when a massive landslide from Xiaolin Village in Kaohsiung County buried 462 residents under a sea of mud. The government's postdisaster relocation policy for Indigenous communities, particularly in Namasia District of Kaohsiung, sparked divisions within communities and widespread public controversy.

The impact of Typhoon Morakot was profound, exposing the increasing human vulnerability to extreme weather events probably caused by climate change. Developmentalist policies implemented in the 1960s promoted road construction through the fragile Central Mountain Range and investment in agriculture and forestry in southern mountainous areas. Indigenous communities in these regions had already been displaced multiple times, exacerbating both their ecological and social vulnerabilities. Because the region had not experienced a perfect storm for a long time, the compounded risks of environmental degradation and climate change remained largely unnoticed (Morakot Independent News Network, 2013). Climate change has altered the frequency and intensity of extreme weather events. Historically, Taiwan experienced an average of three or more typhoon landfalls per year. Climate models developed by the IPCC for southern Taiwan indicate that, due to global warming, intense typhoons are becoming less frequent but bring greater rainfall and more intense storms. Despite heavy government investment in water infrastructure, existing flood control and warning systems have proven to be inadequate against sudden extreme rainfall and the deadly landslides (Lin et al., 2022).

Beyond landslides, Typhoon Morakot also caused downstream flooding, affecting wetlands near river estuaries in southwestern Taiwan. These low-cost lands had seen significant depopulation, leading to two major vulnerabilities. First, fishers expanded the scale of large-scale fish farming, excessively extracting groundwater, causing land subsidence and accelerated coastal erosion (Chen et al., 2022). Second, many abandoned village buildings have been converted into nursing homes for the elderly, many of whom are immobile. The typhoon-induced floods claimed casualties in some of these care facilities, prompting government rescue operations by firefighters and the military—a response that faced harsh public criticism (Hung & Lu, 2015).

Following the 1999 Earthquake, Taiwan established the National Science and Technology Center for Disaster Reduction to conduct scientific disaster research. After Typhoon Morakot in 2010, a nationwide survey tracked 1,754 displaced households over the next decade, making it one of the most comprehensive databases on climate-induced displacement worldwide (National Science and Technology Center for Disaster Reduction, 2025). Additionally, Taiwan's Central Weather Administration made its Typhoon Database public (Chung et al., 2022), and the Soil and Water Conservation Bureau developed a Debris Flow Prevention Information Network, both of which have become critical scientific resources for research on typhoons and climate change (Agency of Rural Development and Soil and Water Conservation, 2025).

Using these datasets, I analysed the natural and social risk factors of the Morakot disaster. My findings revealed that higher landslide risk and total rainfall were correlated with greater loss in economic value and human health. Even after natural hazard factors were controlled for, social inequality was found to significantly influence disaster vulnerability. Similar to findings from earthquake studies, socioeconomic disparities played a crucial role, with farmers, blue-collar workers and unemployed individuals facing disproportionately high disaster risks. Indigenous families were particularly vulnerable, experiencing a disaster risk six times higher than that of their counterparts in the general Taiwanese population because of their overrepresentation among blue-collar workers and farmers, whose occupations carry more hazards than others. Nevertheless, Indigenous communities also demonstrated stronger social capital, which facilitated faster income recovery. Households with greater financial losses and higher injury rates were more likely to receive financial support from friends and relatives, and Indigenous identity, along with frequent interactions with neighbours and communities, was positively associated with higher levels of postdisaster assistance (Lin et al., 2017).

The post-Morakot period saw heightened tensions between the state and civil society, leading to institutional innovations that sparked considerable controversy. The government, in collaboration with major religious organisations, implemented a large-scale, top-down relocation policy, moving Indigenous communities from landslide-prone areas to nearby flatlands owned by the state. Although they were granted usage rights for housing, they were not given land ownership, and many resettlements were located far from their original agricultural lands. Consequently, some villagers returned to disaster-prone areas, and others migrated to nearby cities, leading to significant criticism of what was perceived to be a technocratic policy that lacked sufficient community input (Kao & Lu, 2021). Additionally, legal amendments granted the government the authority to forcibly evacuate residents before an approaching typhoon, with a particular focus on the early evacuation of nursing home residents in the storm's projected path. Although this policy raised human rights concerns, it proved highly effective in preventing major casualties in subsequent typhoons (Ma, 2012). One unexpected outcome of Typhoon Morakot was a sharp increase in public awareness of climate change risks in Taiwan. Since then, climate scepticism in the country has remained at less than 10% in public opinion polls, significantly contributing to government and civil society efforts in promoting renewable energy initiatives (Lin et al., 2022).

Compared with the Chi-Chi Earthquake, postdisaster state–society relations following Typhoon Morakot were more imbalanced. The government implemented a top-down village relocation programme, which encountered strong resistance from Indigenous communities. Our research shows that communities which protested and compelled the government to hold public coordination meetings experienced slower reconstruction progress. However, in the long term, these communities demonstrated stronger social cohesion and better mental health outcomes. Thus, the post-Morakot reconstruction process adds another case to the Red Queen Effect—illustrating how the dynamic interplay between state authority and civil society resistance can shape institutional resilience in disaster recovery.

## 5 Epidemics: from SARS to COVID-19

In late December 2019, Dr. Ai Fen, director of the emergency department at Wuhan Central Hospital, received a test report confirming that staff at the Huanan Seafood Market had contracted SARS. She immediately took a photo of the report and shared it in a WeChat group of medical colleagues, warning them about the situation and informing her superiors. In 2003, SARS had spread rapidly through China, Hong Kong, and Taiwan, with a mortality rate exceeding 10%. The Wuhan Municipal Health Commission issued an emergency notice about an outbreak of pneumonia of unknown cause. The return of a deadly coronavirus alarmed the medical community. One of Ai Fen's colleagues, ophthalmologist Li Wenliang, recognised the severity of the situation and forwarded the report to the public on Sina Weibo, a Chinese social media. The Wuhan government, seeking to suppress media exposure, ordered Ai Fen and Li Wenliang not to spread any further information. The city's health commission later reported 41 cases of 'viral pneumonia of unknown cause', mostly concentrated in the seafood market, and falsely claimed there was no evidence of human-to-human transmission or infections among healthcare workers. Tragically, Li Wenliang, now regarded as a whistleblower, contracted the virus on January 10 and died a month later (Sheng, 2020).

With Wuhan officials suppressing information and Beijing responding sluggishly, the virus quickly spread beyond China's borders. By January 13, 2020, cases had been detected in Thailand, Japan, and South Korea, and by January 21, infections appeared in the US city of Seattle. By February, China reached the peak of its outbreak, and cases in Italy, South Korea, and Iran surged. On March 11, as infections spread rapidly across the United States, Europe, and the Middle East and began reaching Latin America and Africa, the World Health Organization, after weeks of delay and Beijing's attempts to downplay the situation, finally declared COVID-19 a global pandemic. Over the next 3 years, the pandemic led to an estimated 18 to 35 million deaths, making it the most catastrophic disaster of the 21st century (The Economist, 2022).

Remarkably, despite its close economic and demographic ties to China, Taiwan implemented some of the most effective pandemic response measures globally. In the first year of the outbreak, Taiwan prevented large-scale community transmission, with confirmed infections largely limited to travellers entering the country. By the end of 2020, Taiwan had recorded only about 1,200 COVID-19 cases and just 12 deaths. In 2021, the Alpha (B1.1.7) variant triggered a more severe outbreak and spread across the island, resulting in approximately 15,000 infections and 896 deaths. In 2022, after lifting lockdown measures following mass vaccination, Taiwan experienced its most significant wave, with 14,667 fatalities. Despite this, cross-national comparisons ranked Taiwan's pandemic response among the top 10 globally (Keck & Holz, 2021).

Taiwan's success in controlling COVID-19 was a combination of both fortunate circumstances and historical institutional factors. One key advantage was early intelligence from whistleblowers in China, which garnered heightened attention in Taiwan due to its experience with the 2003 SARS outbreak.

On December 31, 2019, Taiwan's Centers for Disease Control sent an email to Beijing inquiring about the outbreak and simultaneously alerted the World Health Organization. This email became evidence that Taiwan had informed the World Health Organization about the Wuhan outbreak before China officially reported it (Lin, 2022).

Ironically, political hostility between Beijing and Taipei helped mitigate the pandemic's initial impact in Taiwan. Prior to the outbreak, China's political sanctions cut monthly Chinese tourist arrivals to Taiwan from nearly 400,000 to just 90,000. On January 12, just after Taiwan's presidential election, the ROC government dispatched two physicians to Wuhan for an on-site investigation. Upon their return on January 15, they reported suspicions that the Chinese Communist Party (CCP government was concealing the true severity of the outbreak. That same day, 6 days before Beijing did, Taiwan officially declared COVID-19 a significant infectious disease. On January 20, recognising the risk of a second SARS-like pandemic, Taiwan established the Central Epidemic Command Center. By January 23, Taiwan confirmed its first domestic case, and on January 26, it banned entry to travellers from Hubei province. As the Wuhan outbreak escalated, Taiwan swiftly suspended all entry permits to travellers from China and Hong Kong and enforced strict quarantine procedures on returning Taiwanese citizens (Lin, 2022).

Taiwan's pandemic response was further strengthened by leaders with public health expertise. Former Minister of Health Chen Chien-jen, who later became vice president, had played a crucial role in Taiwan's response to the 2003 SARS outbreak. He had established protocols for border control, contact tracing, mandatory quarantine, and wage compensation schemes for isolated individuals and promoted preventative measures such as mask-wearing, temperature checks, and hand hygiene. Notably, Vice President-elect Lai Ching-te in 2020, who later became Taiwan's president in 2024, was also a public health expert. This expertise set Taiwan's leadership apart from other democracies. Under their guidance, the Taiwanese government quickly banned the export of masks and provided subsidies to scale up domestic mask and medical supply production, leveraging its renowned manufacturing sector. Public health studies later confirmed that Taiwan's strict border controls, mandatory quarantine measures, and digital contact tracing were highly effective. Widespread mask-wearing and individual compliance with public health measures also contributed significantly to Taiwan's success in pandemic prevention (Ng et al., 2021).

Taiwan's institutional strengths further reinforced its pandemic response. First, Taiwan's universal health insurance system, established in 1995 during the Lee Teng-hui administration, ensured comprehensive health coverage for nearly the entire population and provided an extensive medical data infrastructure (Wong, 2004). Second, Taiwan's experience with SARS in 2003 left lasting lessons, making both health authorities and civil society highly vigilant about Chinese-origin epidemics (Griffiths, 2020). Third, Taiwan's Communicable Disease Control Act underwent major revisions after SARS, empowering the Central Epidemic Command Center with professional leadership. A decade later, in 2013, the government restructured its health administration, creating the Ministry of Health and Welfare and the Taiwan Centers for Disease Control, further enhancing the professionalism and political authority of the Central Epidemic Command Center. Compared with other countries, Taiwan's long history of combating epidemics, combined with its robust healthcare system developed after democratisation, provided its exceptional state capacity to manage infectious disease outbreaks.

While Taiwan's pandemic response was widely praised in the early stages of COVID-19, several areas merit critical reflection. The state's top-down approach to border control and mandatory isolation proved highly effective at containing the virus initially, but it also imposed severe costs on local economies and small- to medium-sized enterprises. Prolonged periods of isolation and social distancing contributed to rising social disconnection and the weakening of civil society organisations. As a result, public support for the ruling Democratic Progressive Party (DPP) declined, contributing to a political deadlock following the 2024 elections in which the executive and legislature came under the control of opposing parties. Compounding this political gridlock, Taiwan now faces an escalating threat of military aggression. These developments underscore the need to balance institutional efficiency with democratic legitimacy and civic participation in times of crisis.

## 6 Geopolitics: the Risk of War

The COVID-19 pandemic brought unexpected consequences for China–Taiwan relations. Beijing refused to take responsibility for its concealment of critical information, instead adopting aggressive 'wolf warrior diplomacy' to deflect blame. Meanwhile, the surge in demand for consumer electronics and semiconductors during the pandemic highlighted Taiwan's critical role in the global supply chain, elevating its strategic influence (Miller, 2022). In 2021, when Taiwan faced a domestic outbreak and a shortage of vaccines, Beijing blocked German pharmaceutical company BioNTech from exporting vaccines to Taiwan, further drawing global attention. This move prompted Japan and the United States to prioritise donations of other vaccines to Taiwan. Despite

Beijing's intensified political pressure on Taipei during the pandemic, Taiwan gained stronger international support. As China's diplomatic standing worsened, Beijing increasingly relied on military intimidation toward Taiwan. This behavioural pattern peaked during Nancy Pelosi's visit to Taipei in 2022, Tsai Ing-wen's visit to the United States in 2023, and Xi Jinping's reappointment as General Secretary of the CCP and President of the PRC in the same period.

Since 1950, sporadic armed conflicts between China and Taiwan have resulted in significant casualties, second only to epidemics, earthquakes, and floods. After the Korean War, Mao Zedong turned his attention to the coastal islands controlled by Chiang Kai-shek, particularly Kinmen, Yijiangshan, and the Dachen Islands. Between late 1954 and early 1955, the People's Liberation Army (PLA) launched attacks on these islands, triggering the First Taiwan Strait Crisis. More than 500 Taiwanese troops were killed. When the U.S. Navy assisted the Nationalist forces in evacuating Yijiangshan and Dachen, Mao declared a ceasefire after securing these territories ("Taiwan Strait Crisis," 2019). In August 1958, the PLA launched a heavy bombardment on Kinmen, sparking major aerial confrontations in what became known as the Second Taiwan Strait Crisis, resulting in more than 600 Taiwanese military casualties. Recent research suggests that this crisis was largely a demonstration of Mao's dissatisfaction with Soviet leader Nikita Khrushchev's policies, further straining Sino–American and Sino–Soviet relations. For more than a decade, China continued regular shelling from Xiamen to Kinmen, causing persistent economic distress in the region (Chen, 2001).

The global geopolitical landscape shifted when U.S. President Richard Nixon met with Mao Zedong in Beijing in 1972. However, China's Taiwan policy remained centred on the 'military liberation of Taiwan' until 1979. On January 1, 1979, when the United States officially established diplomatic relations with China, Beijing ceased its shelling of Kinmen and introduced the 'peaceful reunification, one country, two systems' policy. Initially targeted at Taiwan, this framework was later applied to Hong Kong and Macau. In early February 1979, Deng Xiaoping launched the Sino–Vietnamese War, a seemingly inconclusive conflict that may have been an attempt to improve United States–China relations. This was China's latest formally declared war (Harding, 1992). Soon after, the United States ended its Mutual Defense Treaty with the ROC and withdrew its military from Taiwan, replacing it with the Taiwan Relations Act to indirectly deter military conflict. Meanwhile, China initiated economic reforms and opened its markets to the world (Hass et al., 2023).

The Third Taiwan Strait Crisis did not lead to actual combat but significantly escalated tensions. In 1995, then-President Lee Teng-hui, who was campaigning for Taiwan's first direct presidential election, delivered a speech at his alma mater, Cornell University, signalling United States support for Taiwan's democratic transition. This alarmed the CCP, which had violently suppressed the 1989 pro-democracy movement and was preparing to take over Hong Kong in 1997. In response, China conducted large-scale military exercises and, between July 1995 and March 1996, fired at least 11 missiles into waters near Taiwan's main ports. The United States deployed the aircraft carriers USS Nimitz and USS Independence near the Taiwan Strait to demonstrate support for Taiwan's democracy and security (Hass et al., 2023).

Following Russia's invasion of Ukraine in February 2022, global media raised the question of whether China would attempt a military invasion of Taiwan and whether the United States would intervene. On August 2, 2022, U.S. House Speaker Nancy Pelosi visited Taiwan with five congressional representatives, meeting President Tsai Ing-wen as a diplomatic and military signal of deterrence to China. On August 4, the PLA launched live-fire military exercises in waters and airspace surrounding Taiwan. That afternoon, China's Rocket Force fired 11 missiles into waters near Taipei, southern Taiwan, and the eastern coast—some believed to be nuclear-capable. Japan's Ministry of Defense confirmed that nine of these missiles were detected, five landing in Japan's exclusive economic zone and four flying over Taiwan's main cities. Over the next 4 days, China's Eastern Theater Command continued joint military drills, simulating antisubmarine operations, naval assaults, and amphibious landings. Some analysts labelled this escalation the 'Fourth Taiwan Strait Crisis' (Chen et al., 2025).

China's military intimidation prompted decisive action from Taiwan. In December 2022, then-President Tsai Ing-wen announced the extension of mandatory military service from 4 months to 1 year. Beijing responded by deploying a record number of military aircraft to harass Taiwan's airspace and launching large-scale drills in January 2023 (Wang, 2023). In late March 2023, Tsai embarked on a diplomatic tour to Guatemala and Belize, transiting through the United States. On March 30, she arrived in New York and received the 'Global Leadership' award from the Hudson Institute. On April 5, she delivered a public speech at the Reagan Library in California, where she met with U.S. House Speaker Kevin McCarthy, who referred to her as the 'President of Taiwan'. Meanwhile, the PLA conducted another round of military exercises, sending the Shandong aircraft carrier near Taiwan as a show of force. These drills approached historic levels of military activity (Up Media, 2023).

The threat of war stimulates state building on military force. To counter China's growing threat, Taiwan implemented significant defence measures in 2024, including the extension of conscription to one year and a steady increase in defence spending from 2% to 3% of gross domestic product between 2018 and

2025. Despite these measures, Taiwan's military capabilities remain dwarfed by those of the People's Liberation Army (PLA), and persistent personnel and equipment shortages highlight the urgent need for comprehensive training reforms. Taiwan must draw lessons from Ukraine's defence against Russia, particularly in adopting asymmetric and high-tech defence strategies—such as the deployment of drones and precision-guided systems—to deter and respond to potential invasions (Heraud, 2024).

On the other hand, institutional resilience also depends on an active civil society. Civil society groups in Taiwan have taken proactive steps, inspired by Ukraine's grassroots mobilisation, to support national defence through decentralised and community-based initiatives. Organisations such as Black Bear Academy and Forward Alliance, both youth-driven platforms, and the Presbyterian Church in Taiwan's "Ark Project," have developed training programmes for civil defence and disaster preparedness. These initiatives aim to integrate civil society into wartime logistics, local preparedness, and emergency response planning. In conclusion, Beijing's potential invasion may constitute the ultimate test of Taiwan's institutional resilience in the foreseeable future.

## 7 Conclusion: the Secret of the Red Queen

Taiwan serves as a case study illustrating the intersectionality of risks and institutional resilience in modern society. Due to its geographic location and geophysical characteristics, Taiwan faces multiple overlapping risks, namely the 'Four Horsemen' of earthquakes, typhoons, pandemics, and war. Even more concerning is that these risks often compound each another, leading to severe losses in human wellbeing. One of the most common interactions is that between earthquakes and typhoons, which trigger devastating landslides. For instance, in the aftermath of the 1999 Chi-Chi Earthquake, landslides during Typhoon Toraji in 2001 caused nearly 300 deaths. Similarly, military conflicts have historically been intertwined with disease outbreaks. The Japanese occupation of Taiwan in 1895 and the KMT's suppression of uprisings after taking control in 1947 both coincided with mass population movements and widespread epidemics, each resulting in more than 20,000 deaths. Taiwan, therefore, provides a critical case for studying the intersectionality of multiple risks.

Despite the violent threats posed by these 'Four Horsemen', Taiwanese society has developed institutional mechanisms for self-preservation. Drawing from Daron Acemoglu and James A. Robinson's 'Red Queen Hypothesis' on the interaction between state capacity and civil society, I apply the concept to institutional evolution in response to disasters (Acemoglu & Robinson, 2019). The Red Queen effect, in this context, describes the competitive yet adaptive dynamic between state and civil society's capacity for self-organisation. In the wake of successive disasters, this effect enables societies to internalise past traumatic experiences and strengthen institutional defences against future crises. I refer to this ability as institutional resilience.

Over the past 3 decades, Taiwan has experienced a series of major disasters, some of which shattered half-century records for casualties and destruction, thereby accelerating the Red Queen Effect. The 1999 Chi-Chi Earthquake, the second deadliest earthquake in Taiwan's history, claimed 2,444 lives. Occurring during Taiwan's democratic transition, the earthquake prompted swift legislative action, leading to the enactment of the Disaster Prevention and Protection Act, which authorised both military and civil society participation in disaster relief and reconstruction. The 2009 Typhoon Morakot disaster and resulting landslides killed 699 people and forced Taiwan to reassess its development-driven environmental policies and acknowledge the vulnerabilities and rights of Indigenous communities. The disaster also led to the government's expanded authority in mandatory evacuations and emergency resettlement, particularly heightening public awareness of climate change. The 2003 SARS outbreak, transmitted from China, prompted Taiwan to reform its Communicable Disease Control Act, strengthening the Centers for Disease Control and establishing the Central Epidemic Command Center. These institutions proved critical during the 2020 COVID-19 pandemic, significantly reducing Taiwan's infection and mortality rates despite substantial social disruptions. Taiwan's sociological research on disasters has primarily focused on these three crises, highlighting not only the intersection of multiple risks and social inequalities but also documenting the institutional adaptations that emerged in their aftermath (Chang & Lin, 2012; Lin et al., 2018; Lin, 2022; Lin et al., 2022).

Past disaster experiences and institutional resilience—particularly the Red Queen effect, which reflects the competition and collaboration between state capacity and civil society mobilisation—may determine Taiwan's ability to withstand future risk intersections. Indeed, over the past 70 years, Taiwan's mass-casualty events have shifted from armed conflict to primarily natural disasters and pandemics, mirroring a peaceful era in East Asia from the 1980s. Nevertheless, catastrophic events and the looming threat of military invasion continue to pose significant risks to the wellbeing of its people. Armed conflicts can also intersect with other crises, such as pandemics and natural disasters. As a result, this complex web of interwoven risks will remain a continuous challenge for institutional resilience in both Taiwan and the broader East Asian region.

## Bibliography

Acemoglu, D., & Robinson, J.A. (2019). *The narrow corridor: States, societies, and the fate of liberty*. Penguin Press.

Agency of Rural Development and Soil and Water Conservation (ARDSWC). (2025). *Debris Flow and Large-Scale Landslide Disaster Prevention Information*. https://246.ardswc.gov.tw/?lang=en.

Andrade, T. (2008). *How Taiwan became Chinese: Dutch, Spanish and Han colonization in the seventeenth century*. Columbia University Press.

Centre for Research on the Epidemiology of Disasters (CRED). (2025). *EM-DAT: The International Disaster Database*. https://www.emdat.be/.

Chang, Y.-C., & Lin, T.-H. (2012). Unequal disaster: Hazard and social stratification in an earthquake. *Journal of Social Sciences and Philosophy*, *24*(2), 193–231. https://doi.org/10.6350/JSSP.201206.0193 [In Chinese].

Chao, W.-I., & Lin, T.-H. (2019). Social capital, institutional innovation, and condominium apartments rebuilding of the 1999 Chi-Chi earthquake in Taiwan. *Formosa Journal of Mental Health*, *32*(4), 367–405. https://doi.org/10.30074/FJMH.201912_32(4).0001 [In Chinese].

Chen, J. (2001). *Mao's China and the Cold War* (pp. 163–204). The University of North Carolina Press.

Chen, K.-C. (1992). A demographic estimate of deaths in the February 28 Incident. In EXECUTIVE YUAN RESEARCH GROUP ON THE FEBRUARY 28 INCIDENT (Ed.), *Research report on the February 28 Incident* (Appendix 5). Executive Yuan. [In Chinese].

Chen, L.-C. (2022). Institutional resilience: Institutional reform and technological foresight in Taiwan's disaster risk governance. In T.-H. Lin et al. (Eds.), *Rebirth from the quake: Scientific investigations of risk and institutional resilience after the 1999 Chi-Chi earthquake in Taiwan* (chap. 10). National Taiwan University Press. [In Chinese].

Chen, W.-K. (2019). Study on the battle of the sea and air in the 'Battle of Guningtou'. *Navy Professional Journal*, *53*(6), 6–22. https://doi.org/10.6237/NPJ.201912_53(6).0001. [In Chinese].

Chen, X.-Y., et al. (2025). Deterrence, assurance, and countermeasures: Insights from Pelosi's Taiwan visit and Chinese military exercises. *Asian Survey*, *65*(1), 131–159. https://doi.org/10.1525/as.2024.2408270.

Chen, Y., et al. (2022). Assessment of flood risk map under climate change RCP8.5 scenarios in Taiwan. *Water*, 14(2), 207. https://doi.org/10.3390/w14020207.

Cheng, S.-N., et al. (1999). *Atlas of the ten major earthquake disasters in Taiwan*. Central Weather Administration and Institute of Earth Sciences, Academia Sinica. [In Chinese].

Cheng, W. (2021). *War, trade and piracy in the China Seas (1622–1683)*. Acropolis. [In Chinese].

Chung, L.-L., et al. (2022). Earthquakes don't kill, buildings do: Seismic risk and resilience in architecture. In T.-H. Lin et al. (Eds.), *Rebirth from the quake: Scientific investigations of risk and institutional resilience after the 1999 Chi-Chi earthquake in Taiwan* (chap. 4). National Taiwan University Press. [In Chinese].

Fan, Y. (1995). Plague and public health in Taiwan (1896–1917). *Bulletin of the Taiwan Branch Library, 1*(3), 59–84. [In Chinese].

Fan, Y. (1996). Medicine and colonial expansion: Taiwan's malaria research under Japanese rule. *New History, 7*(3), 133–173. https://tpl.ncl.edu.tw/NclService/JournalContentDetail?SysId=A96022330 [In Chinese].

Government-General of Taiwan. (2025). *Vital statistics of Taiwan (1914–1943)*. Office of the Chief of the Civil Affairs Department. https://rarebooks.ith.sinica.edu.tw/sinicafrsFront99/search/search_result.htm [In Japanese].

Griffiths, J. (2020, April 5). *Taiwan's coronavirus response is among the best globally*. CNN. https://edition.cnn.com/2020/04/04/asia/taiwan-coronavirus-response-who-intl-hnk/index.html.

Hass, R., et al. (2023). *U.S.-Taiwan relations: Will China's challenge lead to a crisis?* Rowman & Littlefield Publishers.

Harding, H. (1992). *A fragile relationship: The United States and China since 1972* (pp. 215–246). The Brookings Institution.

Heinrich, V.F. (2010). *What makes civil society strong* (Doctoral dissertation). Fernuniversität Hagen.

Hendrix, C.S. (2010). Measuring state capacity: Theoretical and empirical implications for the study of civil conflict. *Journal of Peace Research, 47*(3), 273–285. https://www.jstor.org/stable/20752162.

Heraud, R. (2024). *The new 12-month conscription programme in Taiwan as a catalyst for the transformation of civil-military relations: Military education and Generation Z* [Taiwan Fellowship 2023 Report, with additional Traditional Chinese translation by Jen-feng Kuo]. https://taiwanfellowship.ncl.edu.tw/eng/scholar_detail.aspx?sn=96764.

Huang, S.-M. (2011). A historical study of the 1959 flood disaster. *Hualien Historical Journal, 8*, 61–89. https://doi.org/10.29729/NDHUSJHR.201110.0004 [In Chinese].

Hung, H.-C., & Lu, Y.-T. (2015). Adaptation to climate change and typhoon-flood hazards in coastal areas. *City and Planning*, 42(1), 87–108. https://doi.org/10.6128/CP.42.1.87.

Intergovernmental Panel on Climate Change (IPCC). (2025). *Managing the risks of extreme events and disasters to advance climate change adaptation*. https://www.ipcc.ch/report/managing-the-risks-of-extreme-events-and-disasters-to-advance-climate-change-adaptation/.

Ka, C.-M. (2021). *Cultivated aborigines and the unruled Hans: Governance deployment and contentious politics in Qing Taiwan*. National Taiwan University Press. [In Chinese].

Kao, R.-L., & Lu, J.-C. (2021). Housing recovery and free-permanent housing policy assessment after Typhoon Morakot in Taiwan. *City and Planning, 48*(4), 436–461. https://doi.org/10.6128/CP.202112_48(4).0004 [In Chinese].

Keck, M., & Holz, N. (2021, February 2). New Zealand, Vietnam, Taiwan top list of countries with the most effective COVID-19 responses. *Global Citizen*. https://www.globalcitizen.org/en/content/new-zealand-vietnam-top-list-covid-19/.

Khan, S.W. (2024). *The struggle for Taiwan: A history of America, China, and the island caught between*. Basic Books.

Lai, C.-H. (2015). *Epidemics and the February 28 Incident* (pp. 120–121). Wu San Lien Taiwan Historical Materials Foundation. [In Chinese].

Lin, K.-H., & Chen, T.-H. (2022). The victimization of the vulnerable: Social vulnerability in Taiwan's seismic risk. In T.-H. Lin et al. (Eds.), *Rebirth from the quake: Scientific investigations of risk and institutional resilience after the 1999 Chi-Chi earthquake in Taiwan* (chap. 5). National Taiwan University Press. [In Chinese].

Lin, K.-H.E., et al. (2015). An interdisciplinary perspective on social and physical determinants of seismic risk. *Natural Hazards and Earth System Sciences, 15*, 2173–2182. https://doi.org/10.5194/nhess-15-2173-2015.

Lin, K.-H.E., et al. (2017). How does resilience matter? An empirical verification of the relationships between resilience and vulnerability. *Natural Hazards, 88*(2), 1229–1250. https://doi.org/10.1007/s11069-017-2916-1.

Lin, L.-Y., et al. (2022). The Taiwan climate change projection information and adaptation knowledge platform: A decade of climate research. *Water, 14*(3), 358. https://dx.doi.org/10.3390/w14030358.

Lin, T.-H. (2015). Governing natural disasters: State capacity, democracy, and human vulnerability. *Social Forces, 93*(3), 1267–1300. https://doi.org/10.1093/sf/sou104.

Lin, T.-H. (2022). An initial exploration of risk governance in Taiwan's COVID-19 pandemic: Historical opportunity, exposure, vulnerability, and resilience. *Reflexion, 44*, 233–266. https://www.airitilibrary.com/Article/Detail?DocID=a0000477-202201-202204120015-202204120015-233-266 [In Chinese].

Lin, T.-H. (forthcoming). Disasters and recovery. In *Encyclopedia of Taiwan Studies Online*.

Lin, T.-H., et al. (Eds.). (2022). *Rebirth from the quake: Scientific investigations of risk and institutional resilience after the 1999 Chi-Chi earthquake in Taiwan*. National Taiwan University Press. [In Chinese].

Lin, T.-H., et al. (2018). Approaching a world risk society? Social capital, risk perception and risk coping behavior in Taiwan. *Survey Research—Methods and Application, 40*, 127–166. https://doi.org/10.7014/SRMA.201809003 [In Chinese].

Lin, T.-H., et al. (2022). *Sociology of carbon reduction: Public opinion and challenges in Taiwan's net-zero transition*. Chuliu Book Co. [In Chinese].

Lo, M.-C.M., Lin, T.-H., & Lin, W.-Y. (Eds.). (2024). *Taiwan's COVID-19 experience: Governance, governmentality, and the global pandemic*. Routledge.

Ma, K.-C., et al. (2012). Evaluation of the effect of hysteretic flow and root system on shallow landslide. *Soil Research, 50*, 616–624. https://doi.org/10.1071/SR12104.

Miller, C. (2022). *Chip war: The fight for the world's most critical technology*. Scribner.

Morakot Independent News Network. (2013). *Thinking of home in permanent housing: Three years after Morakot, permanent housing policy and stories*. https://www.88news.org/posts/category/columns/%E6%96%B0%E5%88%86%E9%A1%9E%E5%90%8D%E7%A8%B1%E5%9C%A8%E6%B0%B8%E4%B9%85%E6%83%B3%E5%AE%B6%E2%94%80%E6%B0%B8%E4%B9%85%E5%B1%8B%E6%94%BF%E7%AD%96%E8%88%87%E4%BA%BA%E7%9A%84%E6%95%85%E4%BA%8B [In Chinese].

National Science and Technology Center for Disaster Reduction. (2025, March 17). *Social impacts and recovery survey of Typhoon Morakot (2010–2019)*. https://drrstat.ncdr.nat.gov.tw/statistics/morakot-en.

Ng, T., et al. (2021). Comparison of estimated effectiveness of case-based and population-based interventions on COVID-19 containment in Taiwan. *JAMA Internal Medicine, 181*(7), 913–921. https://doi.org/10.1001/jamainternmed.2021.1644.

Özerdem, A., & Jacoby, T. (2006). *Disaster management and civil society: Earthquake relief in Japan, Turkey and India*. I.B. Tauris.

Oronce, C.I.A., et al. (2025). *Association Between State-Level Income Inequality and COVID-19 Cases and Mortality in the USA*. https://doi.org/10.1007/s11606-020-05971-3.

Putnam, R.D., Leonardi, R., & Nanetti, R.Y. (1993). *Making democracy work: Civic traditions in modern Italy*. Princeton University Press.

Sheng, Y. (2020, March 11). COVID-19 pandemic: The 'whistleblower' sparked an anti-censorship battle, Chinese netizens creatively fought back. *BBC Chinese News*. https://www.bbc.com/zhongwen/trad/chinese-news-51831652 [In Chinese].

Skocpol, T. (1985). Bringing the state back in: Strategies of analysis in current research. In P.B. Evans, D. Rueschemeyer, & T. Skocpol (Eds.), *Bringing the state back in* (pp. 3–38). Cambridge University Press.

Taiwan Strait Crisis. (2025, March 17). In *Wikipedia*. https://zh.wikipedia.org/wiki/%E8%87%BA%E7%81%A3%E6%B5%B7%E5%B3%BD%E5%8D%B1%E6%A9%9F#%E7%AC%AC%E4%B8%80%E6%AC%A1%E8%87%BA%E7%81%A3%E6%B5%B7%E5%B3%BD%E5%8D%B1%E6%A9%9F%EF%BC%881954%EF%BC%8D1955%EF%BC%89.

The Economist. (2022, October 25). *The pandemic's true death toll*. https://www.economist.com/graphic-detail/coronavirus-excess-deaths-estimates.

Tilly, C. (1992). *Coercion, capital, and European states, AD 990–1992*. Wiley-Blackwell.

UCDP. (2025). *UCDP/PRIO Armed Conflict Dataset version 23.1.* UCDP Dataset Download Center. https://ucdp.uu.se/downloads/.

UP Media. (2023, April 4). McCarthy announces meeting with Tsai Ing-wen; PLA deploys 23 aircraft and warships near Taiwan. https://reurl.cc/Q4x840 [In Chinese].

Wang, J.-H. (2023, January 9). 2023: The first PLA military drill targeting Taiwan! Ministry of National Defense detected 57 aircraft incursions, Presidential Office strongly condemns. *RW NEWS*. https://reurl.cc/rLAymE [In Chinese].

Weng, C.-Y. (2006). The origins of the name Formosa and the 1582 shipwreck of the Portuguese near Taiwan. *Hanlin Social World, 5*, 4–13. [In Chinese].

Wong, J. (2004). *Healthy democracies: Welfare politics in Taiwan and South Korea.* Cornell University Press.

World Economic Forum. (2025). *Global risks report 2025.* https://reports.weforum.org/docs/WEF_Global_Risks_Report_2025.pdf.

Yanaihara, T. (2022). *Taiwan under Japanese imperialism.* Common Master Press. [In Chinese].

Yang, D.L. (2024). *Wuhan: How the Covid-19 outbreak in China spiraled out of control.* Oxford University Press.

Zhang, Y., et al. (2022). Assessment of flood risk map under climate change RCP8.5 scenarios in Taiwan. *Water, 14*(2), 207. https://doi.org/10.3390/w14020207.

CHAPTER 6

# To Rebuild or Not to Rebuild? Disorientation and Re-grounding in Post-quake on Mount Etna

*Mara Benadusi, Rita Ciccaglione, Mario Mattia and Vincenzo Lo Bartolo*

## 1 Introduction[1]

Là où il a tremblé, il tremblera

This statement by the French scientist Georges-Louis Leclerc de Buffon (1707–1788)— "Where it has trembled, it will tremble again"—succinctly encapsulates the guiding idea behind this chapter. Evoking the recurrence of earthquakes in seismically active areas, the maxim functions as an empirically grounded verdict, marking communities situated along active tectonic structures as perpetually threatened by future seismic events.

In this context, examining cultural frameworks of risk perception within a territory characterised by high seismic hazard takes on strategic importance from an applied research perspective. Tracing a form of "collective sensibility" that bridges both pre- and post-disaster phases may offer valuable insights into how local populations respond to institutional measures such as evacuation, reconstruction, or relocation. At the same time, such inquiry may help reorient risk mitigation policies toward more inclusive and context-sensitive approaches.

Exploring "cultures of risk" (Lupton 2013) within communities fully aware of inhabiting fragile, hazard-prone territories is a well-established topic in both Italian and international anthropological literature on disasters (Benadusi 2015; Gugg 2017; Ciccozzi 2016; Bankoff 2003; Lupton 1999; Douglas 1986, 1992; Douglas & Wildavsky 1983;). Existing research indicates that, in such contexts, attitudes of resignation and fatalism regarding the potential recurrence of destructive events are not uncommon (Barrios 2011; Button 2010).

1 The chapter is the result of a collaborative work but the sections can be distributed as follows: sections 1 and 3 to Mario Mattia; sections 2 and 6 to Mara Benadusi; section 4 to Vincenzo Lo Bartolo; section 5 to Rita Ciccaglione.

HTTPS://DOI.ORG/10.1163/9789004751804-008

In the case of Mount Etna, however, we observe a peculiar phenomenon: The so-called "surface faulting," a deformation caused by the rupture propagating from the seismogenic fault plane deep underground up to the surface. For those most exposed to danger, such deformation is not merely the visible trace of the fault line itself; it also symbolizes a qualitatively distinct relationship with the earth's tremor—one that uniquely affects them, setting their experience apart from that of others. It is this selective manifestation that certain families residing on the volcano's slopes are compelled to confront—families for whom the very presence of the fault profoundly shapes their daily lives.

The territory in question—particularly along its eastern flank—is marked by a specific condition: Frequent recurrence of moderate-magnitude, shallow-focus earthquakes that nonetheless inflict severe damage to both private and public buildings. Alongside this "impulsive" activity (earthquakes), the area experiences a geological phenomenon known as "aseismic creep"—a continuous, slow ground movement along lines of weakness that typically correspond to the volcano's active and "capable" faults (i.e., those that reach the surface). This ongoing movement also causes considerable damage to the built environment and, together with earthquakes, highlights the urgent need to explore all possible solutions aimed at mitigating its primary adverse effect: Harm to homes and public infrastructure.

Setting aside strictly technical considerations, this paper presents the preliminary, in-progress findings of a study that seeks to integrate existing geophysical knowledge of the area with socio-anthropological insights. The aim is to explore the condition of disorientation associated with relocation, as well as the process of "settling in" linked to new housing solutions selectively assigned to families affected by the seismic event who had lived directly above the active fault. Given its frequent use throughout this paper, the term *relocation* requires clarification: Here, it refers to the process of transferring family units or entire settlements located in areas subject to recurrent destructive events—particularly of seismic origin—to new residential zones offering reduced exposure to such risks.

This article examines the relocation process that affected 120 families following the December 26, 2018 earthquake, which struck an area extending from the hamlet of Fleri (Zafferana Etnea, CT) to Aci Platani (Acireale, CT). As part of an applied research project, we conducted fieldwork in the area, engaging deeply with nine families to explore the lived experiences of those subjected to selective, state-imposed relocation. In two cases, relocation also involved the displacement of commercial activities housed within the same buildings as the residences. We also interviewed individuals of public roles and actors mediating between citizens and institutions—such as the presidents of

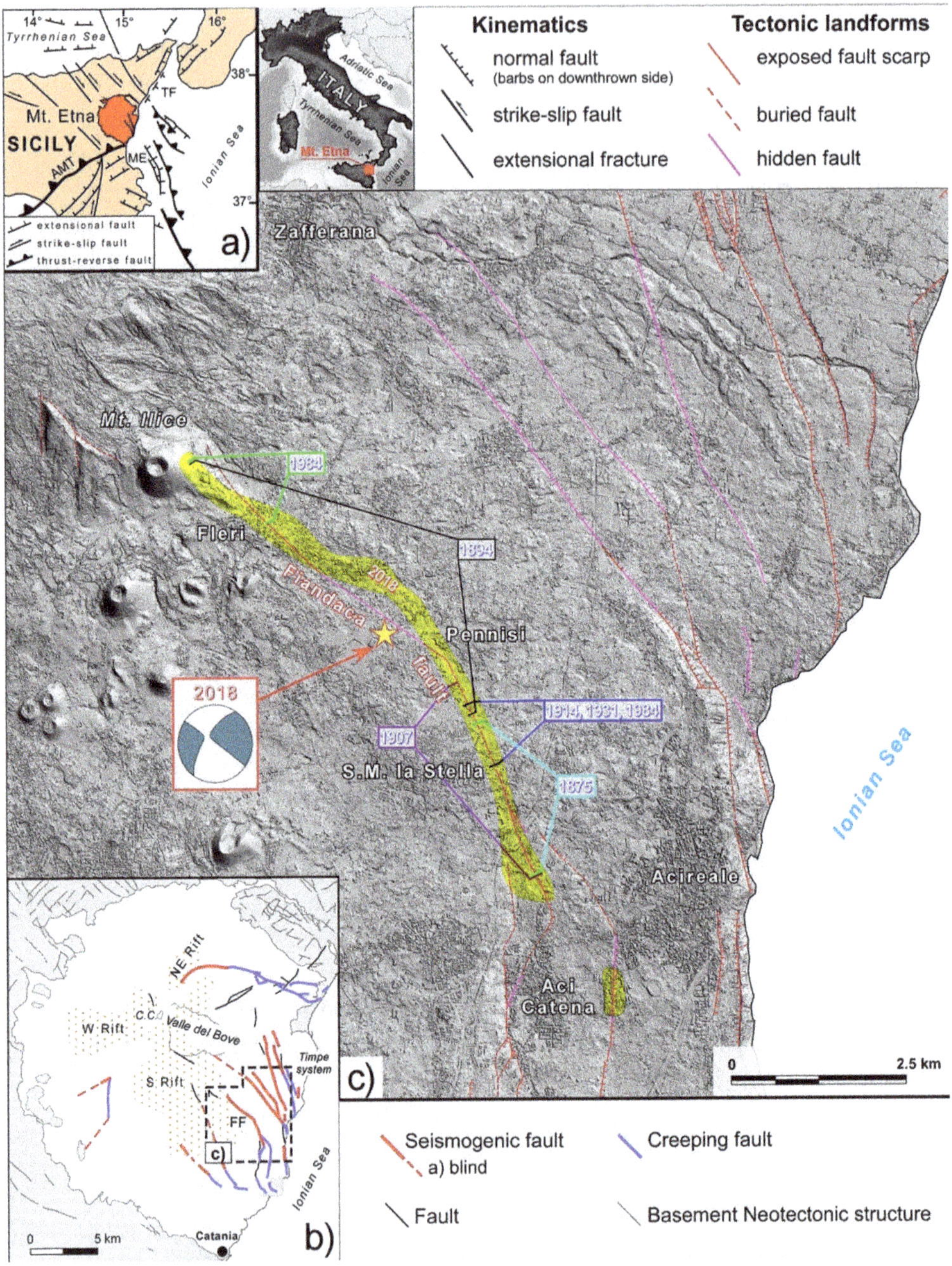


FIGURE 6.1 Tectonic setting of the study area. a) Location map showing Mt. Etna in the framework of the Apennines-Maghrebian thrust-belt; b) sketch of the main active fault-systems on Mt. Etna (FF = Fiandaca Fault); c) detail of the Fiandaca Fault showing the extent of the 2018 surface faulting (yellow band), the focal mechanism of the 26 December 2018 earthquake and the historic surface faulting events

SOURCE: CIVICO ET AL., 2019

committees representing earthquake survivors from two particularly affected hamlets, as well as those responsible for relocation initiatives under the Etnean reconstruction authority.[2] The broader aim of the research was to contribute to the development of more effective public policies for communities living on the slopes of Mount Etna and to offer insights potentially relevant to other territories exposed to similar geophysical risks.

After outlining the theoretical framework that informs our analysis and briefly introducing the research context and methodology, this essay turns to a discussion of the fieldwork findings. The selective relocation process brought into focus a growing awareness of seismic vulnerability. While risk had long been normalized through local knowledge, the earthquake and ensuing displacement unsettled this balance, surfacing underlying fears. Residents engaged more actively with expert knowledge, using it as a means to negotiate agency within the reconstruction process. Institutional trust was reinforced through direct contact with SCRAE officials and the provision of financial incentives, fostering acceptance of resettlement framed as mutually beneficial.

However, official responses remained narrowly focused on the fault zone and private housing, leaving broader recovery—such as public services and local economies—largely peripheral within this fragmented institutional landscape. Slow, uneven reconstruction disrupted social cohesion, particularly among older residents, redirecting attachments toward family networks and emotionally significant belongings rather than community or place. Local committees provided bureaucratic assistance but engaged little in collective advocacy.

Ultimately, the relocation program emerges as a complex process where vulnerability, belonging, and agency are continually negotiated. Rather than passive acceptance or outright resistance, it constituted a form of situated adaptation shaped by experience, memory, and future aspirations, pointing to the need for more nuanced, participatory approaches to disaster risk reduction.

## 2 Risk Perception, Disorientation and Re-grounding of Presence in Relocation Policies

"Earthquakes are frightening." While seemingly obvious—even banal—this statement has far-reaching implications for how populations perceive and respond to the challenges of inhabiting areas marked by high seismic hazard, particularly when compounded by the structural vulnerability of the built environment, as in the Etnean region.

2 Henceforth, we will refer to it using its official acronym, SCRAE.

Recent clinical psychology studies have aimed to quantify earthquake-related fear using specialized measurement scales (Satici et al. 2023), identifying obsessive-compulsive anxiety patterns among affected individuals (Kartol et al. 2024; Neri & Neri 2025). These insights provide a foundation for designing targeted interventions to support psychological recovery after major seismic events. The term *seismophobia* is often used to describe this intense and persistent anxiety triggered by seismic tremors: An anxiety frequently dismissed as excessive or even "irrational." Yet such classifications are not without their limitations.

Living on the slopes of an active volcano or along an active fault line often elicits what anthropologists describe as the "disaster paradox" (Diamond 2005). While fear increases with proximity to the source of danger, once a certain threshold of exposure is crossed, fear paradoxically recedes or disappears altogether. Faced with the prospect of future tremors, people may respond with behaviors that contradict fear—such as forgetting, suppression, or outright denial of the risk.

This dynamic, described by Gugg (2013) as a process of "scotomization,"[3] may appear paradoxical. Indeed, it produces a form of "cognitive invisibility" that conceals danger precisely when it becomes most imminent. In such contexts, bodily experience plays a crucial role. Engaging directly— "with eyes and hands"—with the intimate spaces of one's home and surroundings can reveal subtle signs of risk that might otherwise go unnoticed. Equally important is how individuals engage with the plurality of earthquake-related knowledges—vernacular, devotional, and eschatological—each imbued with rich symbolic meaning.

An anthropological approach that embraces cognitive, emotional, and sociocultural responses to disasters offers crucial insights into how affected communities perceive, experience, and navigate risk, resilience, and recovery processes. For example, the reconstruction model based on "where it was, as it was," used after the 1976 Friuli earthquake, can be seen as a collective effort to reclaim what was lost—an attempt to repair the cultural connections broken by the disaster. Building on this idea, the concept of "fault" takes on a dual meaning: Not only geological but also cultural, symbolizing a rupture in the continuity of time and space experienced by affected communities. To explore

3 The dynamic echo anthropological analyses of how risk can become invisible precisely when it is most immediate. Masco (2004) describes a process of "cognitive erasure" in contexts of chronic environmental threat, where sustained exposure leads to the normalization of danger. Similarly, Barrios (2011, 2017) examines how the repetition of disasters fosters affective and perceptual strategies through which danger becomes familiar—no longer registered as exceptional, and at times scarcely perceived at all.

this discontinuity more deeply, it is helpful to examine the experiential dimensions of *spaesamento* (disorientation) and *appaesamento* (re-grounding).

*Spaesamento* can be understood as the disorienting response to forced displacement from places where social and ecological relationships have taken root over time, helping to shape the horizons of meaning individuals and communities come to perceive as familiar in their everyday life. The experience of *spaesamento* can easily give rise to unsettling emotions or forms of territorial anguish which—in a De Martino's perspective (De Martino 2015, 2023)—relate to a "crisis of presence:" A condition marked not only, or primarily, by the loss of a physical place of belonging, but by the disruption of relational, symbolic, and existential ties to the landscape and social-historical fabric. Adopting De Martino's categories, to exist is always to "make a difference," whereas being unable to do so equates to disappearing or to a weakened presence.[4]

This layered understanding of place and presence further illuminates the symbolic and cultural dimensions involved in post-disaster reconstruction. Engaging with these dimensions in the analysis of *spaesamento* and *appaesamento* is crucial, as shared horizons of meaning play a vital role in legitimizing the emergence of new, collectively inhabited spaces of life after catastrophe. And indeed, these symbolic frameworks are pivotal in shaping the negotiations between local populations, technical experts and political authorities—especially during critical decision-making moments in the reconstruction process.

At times, the impulse to restore ties with the past is offset by dynamics of collective removal, which redirect social tensions toward future-oriented policies—often implicitly framed as forward-looking solutions. This imperative to "make space for the future" may involve severing ties with pre-disaster life. In this process, forgetting—or erasing the material ties connecting people to a disrupted world—can become a necessary step toward reinhabiting the present. The refusal to bear witness to catastrophe—along with its erasure from public or institutional memory—often emerges as a structural feature

4 The concepts of *spaesamento* and *appaesamento* employed in this article draw on Ernesto De Martino's profound theorization of existential and symbolic crisis. In *The End of the World* (2023), Zinn translates *spaesamento* as "disorientation" and *appaesamento* as "re-grounding" or the "reestablishment of presence," describing the process through which individuals or communities reclaim their ability to meaningfully inhabit the world after a profound rupture. As Zinn articulates in her introduction: "De Martino's re-grounding of presence involves restoring meaning and the capacity to act in a world that, under the pressure of extreme events, risks becoming uninhabitable" (De Martino 2023: xvi). Thus, *appaesamento* transcends mere physical return to place; it entails a cultural and symbolic reappropriation of lived space, enabling the reconstitution of presence in a world imperiled by dissolution.

of post-disaster recovery processes (Ullberg 2010; Simpson 2020). In reconstruction scenarios, absence and presence, memory and forgetting, past and future are held in a fragile and often contested balance. What is at stake is not merely the repair of buildings, but the re-grounding (*appaesamento*) of presence in a world rendered temporarily uninhabitable by catastrophe. Not only do survivors confront what has been repressed; those involved in emergency management too inevitably encounter the "territorial anguish"[5] provoked by the earthquake.

Prompted by these reflections, our research team at the Department of Political and Social Sciences, University of Catania (authors of this paper), established a collaboration with the National Institute of Geophysics and Volcanology (INGV) and SCRAE. The project centers on two pivotal research questions: First, to investigate how recovery and adaptation processes following catastrophic events are lived, symbolized, and managed by affected communities; second, from an applied perspective, to devise effective strategies and mediation mechanisms between citizens and institutions aimed at overcoming skepticism, resistance, and distrust towards relocation policies.

In contexts such as the one under study here, where risk is selective, fractures can emerge within the same neighborhood or settlement cluster—between those whose homes lie directly atop the fault line and those living at safer distances. The success of any relocation plan depends not only on its logistical feasibility but also on how adaptation processes unfold within the affected community, especially as diverse social responses to reconstruction policies emerge along the fault. From this standpoint, both the economic viability of relocation and the varied impact of disaster across class-based disparities (Bolin and Stanford 1999) must be considered. Access to public funding, contingent upon relinquishing property in high-risk zones, can facilitate the reestablishment of post-earthquake stability and enhance survivors' trust in emergency management institutions.

There is a significant difference between the selective relocation process analyzed in this chapter and the experiences of mass displacement implemented through broad-based reconstruction policies. In the latter, entire

5 The notion of "territorial anguish" refers to a form of existential distress that arises when the symbolic and affective bond between a community and its territory is fractured. Ernesto De Martino employed this concept to describe the disruption of presence that occurs when the familiar world loses its coherence under the pressure of extreme events. While *spaesamento* and *appaesamento* refer to processes of disorientation and re-grounding, territorial anguish highlights the spatial and collective dimension of such crises—where it is not only the self that is destabilized, but the inhabited world itself. In *The End of the World* (2023), De Martino explores how catastrophic events generate a form of cultural breakdown in which the landscape becomes uninhabitable in both material and symbolic terms.

neighborhoods are relocated, as in the case of the Rione Terra in Pozzuoli during the 1970 bradyseismic crisis (Signorelli 1992), or entire towns, such as the settlements moved following the 1968 Belice Valley earthquake (Petino et al. 2023). It is crucial to recognize that both approaches entail specific vulnerabilities. Mass relocation policies often disrupt the lived experience of housing rooted in communal settings, hindering the reweaving of pre-existing social ties. The logistical pressures of relocating large populations in compressed timeframes frequently result in extended unstable conditions, as exemplified by tent camps and temporary settlements. Conversely, selective relocation may cause earthquake survivors to experience a pervasive sense of isolation, arising from their separation from community and belonging—a rupture that can trigger profound forms of disorientation (*spaesamento*).

Ultimately, both forms of displacement—mass relocation (notably in the form of so-called "new towns") and selective relocation—underscore the need to rethink post-disaster policies in ways that address not only physical safety but also the complexity of the socio-cultural ties shaping experiences of dwelling and re-embedding after catastrophe. While mass displacement has received substantial attention in the academic literature, selective relocation—with its dynamics of isolation and community fragmentation—remains underexplored, despite its significant implications for disaster affected populations.

In the case of the Etna earthquake, at the outset of our project, it was particularly important to explore how separation from an established residential fabric, along with perceptions of unequal treatment, may have influenced relationships with reconstruction authorities. From both an anthropological and institutional perspective, research on selective displacement proved especially insightful, offering a grounded critique of the political tendency to expand risk zones indiscriminately—often driven by incentives tied to non-repayable reconstruction funds.

Unlike previous Italian studies that primarily examined *spaesamento* and *appaesamento* through memory (Moscaritolo 2020; Falconieri, Fichera, Valitutto 2021), our interdisciplinary research—bridging geophysical and anthropological sciences and maintaining active dialogue with local authorities—has introduced an innovative public-oriented approach. Our investigation seeks to enhance meaningful communication and collaboration among citizens, institutions, and the scientific community, ultimately contributing to more responsive and informed disaster governance.

## 3 Conducting Research on the Fault Line

As previously mentioned, the context under analysis is marked by the constant, predictable, and frequent recurrence of moderate-magnitude earthquakes. These events occur at very shallow focal depths and are thus capable of producing spatially limited yet particularly intense damage, as demonstrated by the macroseismic surveys conducted following the December 26, 2018 event along the so-called Fiandaca fault (Azzaro et al 2022). One of the variables considered in this study was the strength of territorial bonds which could help explain resistance to relocation among households located directly on the fault line.

In addition to seismic shocks, active “capable” faults around Etna present an additional hazard through aseismic creep—a gradual ground displacement that silently weakens structures over years before they become uninhabitable. Unchecked urban sprawl onto former farmland has amplified exposure. Periodically, creep accelerates into slow slip events, or “slow earthquakes” (Mattia et al. 2015), which recur every one–two years and can inflict irreparable damage within days or weeks. Although the damage they cause rivals that of sudden earthquakes, local communities often underestimate the risk posed by aseismic creeps. In the absence of institutional mechanisms to compensate for the damage incurred, disaffection or resistance to engaging in constructive dialogue with authorities emerged as critical obstacles to implementing effective prevention policies tailored to the unique challenges of this context.

Addressing the diffuse, long-term, and often invisible everyday forms of catastrophe in such contexts demands more than a detailed examination of the geophysical aspects alone. These phenomena are deeply intertwined with processes of disaster embodiment and temporalization, which can be better understood through an anthropological perspective. While a technical-engineering perspective typically frames catastrophe as the threat posed by destructive physical forces (hazards), anthropology highlights the crucial role of socioeconomic and political-institutional vulnerabilities that shape disaster-affected contexts (Oliver-Smith and Hoffman 1999).

Only by examining historical, social, and ecological factors in their interdependence can we fully grasp the “root causes” of disasters, thereby achieving genuine heuristic insight. These factors reveal which elements exert the greatest influence not only during a disaster’s “incubation phase” but also throughout its ongoing management. Patterns of socioeconomic development, local power dynamics, cultural interpretations of catastrophe, symbolic frameworks, and the political ecology shaping human-environment interactions all informed our decision to expand the investigation beyond the strictly physical, tectonic, and morphological characteristics of the affected areas. Our goal was to incorporate their socio-environmental dimensions, which represent deeply embedded and intimate aspects of the disaster experience.

From this vantage point, the collaboration between geologists and anthropologists proved both fruitful and promising. The succession of microseismic events and the gradual deterioration of buildings due to aseismic creep correspond, in the lived experience of resident families, to a long sequence of small, everyday adjustments—often spontaneous and routine. These fragmented, protracted adaptations contribute to a "normalization" of catastrophe, obscuring the full extent of risk and contributing to its systematic underestimation. Such recurring phenomena, however, should prompt technical experts and policymakers to critically reflect on the widespread tendency to overlook or minimize warning signs and official information concerning seismic hazards. This process of scotomization can have serious consequences, preventing individuals from adopting necessary precautions in the face of disaster. Prior experiences, coupled with inconsistent or ambiguous communication from authorities or the media, can severely erode preparedness, leading to delays in evacuation and heightened vulnerability.

At the outset of our research, we sought to engage families affected by the relocation process initiated under Ordinance No. 18 (December 21, 2020), which prohibited reconstruction within the so-called "Respect Zone", and provided financial assistance for resettlement within the same municipality. Access to these families was made possible through collaboration with the SCRAE, particularly the Vice-Commissioner, whose support enabled us to bypass time-consuming fieldwork preparation activity. This was especially valuable given the wide dispersal of relocated households beyond the restricted area. While we initially hypothesized that institutional endorsement might hinder openness—particularly in exploring the relationship between earthquake survivors and institutions—this concern ultimately proved unfounded. The research team received both positive and negative feedback regarding national, regional, and municipal authorities involved in the reconstruction, even when the team expanded its network through continued presence and direct engagement in the affected area, going well beyond the initial contacts provided by the authorities.

To investigate the processes of *spaesamento* and *appaesamento* among a select group of individuals affected by relocation—following the declaration of uninhabitability of homes damaged by the 2018 earthquake—we focused on the collection of biographical interviews. As a pilot study, the decision to limit the number of participants was intended to ensure greater analytical depth, thereby capturing the complexity and nuanced embodiment of human agency explored through its "spaces of freedom and/or constraining forces" (Comaroff & Comaroff 1992). Focusing on a limited number of family units allowed us to access the intimate spheres where individual and collective decisions take shape—decisions regarding dwelling, territorial rootedness, and relations with institutions—all inevitably shaped by the structural vulnerabilities of the area. This approach enabled a deeper examination of the interplay between geophysical vulnerability and other, less visible forms of vulnerability—dimensions which remain largely underexplored and often poorly understood by the institutions tasked with reconstruction.

The interviewees[6]—seven men and four women from nine distinct households—were compelled to abandon their homes or places of business following the earthquake, as their properties lay in close proximity to, or directly above, the Fiandaca fault trace. Accordingly, the study prioritized exploring not only family histories before relocation but also the personal and collective experiences related to the earthquake. This approach aimed to understand how each participant responded in the critical weeks following the seismic event. Key themes identified for the interviews included the process of forced relocation: Official communication regarding the presence of the fault line on participants' properties; initial reconstruction plans; negotiated housing solutions; and subsequent developments related to these issues.

Among the households in our study, three owned damaged properties in which they did not permanently reside; these served respectively as a parent's residence, a secondary home, or an investment for their children. The remaining households were living in the affected homes at the time of the earthquake and were subsequently relocated. By the time of the interviews, all participants had accessed compensation funds; three were still negotiating the purchase of new properties, while most had already acquired new homes, frequently relocating to larger municipalities. In four instances, these new residences became secondary homes. Of the two businesses owned by interviewees, one was re-established in rented premises within the resettlement municipality—the spouse's hometown—while the other, supplying essential healthcare goods, continued operating locally from a temporary wooden structure. Beyond the spontaneous narratives concerning the earthquake, emergency response,

6 Mr. Salvo, 54, is a civil engineer and secondary school teacher in Construction and Construction Technologies. He lives with his wife and their twin children; born around the time their earthquake-damaged home was being demolished. Mr. Roberto and Mrs. Maria Ester, aged 79 and 81, are retired administrative employees from well-known local firms. They have three adult children, each with their own family. Mr. Antonio, 57, is a building contractor living with his wife; their two sons work in Northern Italy. Mrs. Maria, 48, is an architect and mother to an 8-year-old daughter. Luigi, 43, is a professional photographer. His wife Carla, 40, works as a specialized educational assistant at a local school. They have two children, aged 11 and 4. Dr. Michela, 62, runs a business in the healthcare-commercial sector, assisted by one of her three children, aged 29. Mr. Mirko, 61, works in the restaurant and tourism sector. He is married and has a daughter living abroad. Mr. Sebastiano, 57, lives with his elderly mother. Mr. Alfio, 85, is a retired man with four daughters; one of them, a nurse who lived next door, was relocated with her family.

housing trajectories, and economic recovery, broader socio-economic impacts emerged as prominent themes. This prompted further engagement with collective representatives, notably the presidents of two survivors' committees who acted as intermediaries between the commissioner's office and residents, ensuring continuous communication throughout the reconstruction.

## 4 "The True Quake Began after the Earthquake"

This sentence, spoken by an elderly farmer from the Belice Valley in Italy as he recounted his experience during and after the January 15, 1968 earthquake, directly introduces some of the central themes of our investigation. In his research, Antony Oliver-Smith (1999; 2010) has shown the extent to which post-catastrophe relocations and resettlements affect populations in both the short and long term. Earthquakes of a certain magnitude—as well as surface faulting—can cause severe losses and damage to material assets. Yet dwelling is an experience that can hardly be separated from the "governance of affects" (Barrios 2017; 2008).

Anthropological analysis reveals how emotions—fear, rejection, and attachment to place—shape perceptions of recovery and reconstruction plans, influencing judgments of their effectiveness and attitudes toward living in high-risk areas. The governance of affects grows more complex when aid creates dependency or disaster management reshapes local power dynamics under the guise of economic revitalization (Klein 2007; Gunewardena & Schuller 2008). Following the 2018 earthquake in the Mount Etna region, SCRAE emerged as the authority deciding who must abandon their homes permanently and who instead could rebuild in situ. The introduction of a selectively targeted, non-repayable funding scheme—mirroring the selective nature of the fault—shaped both the themes and narrative frames expressed by interviewees.

Below are key excerpts from the full interviews that illustrate how emotional responses to relocation are profoundly shaped by the scotomization dynamics discussed earlier. Mr. Alfio "becomes beside himself"—in his own words—over the demolition of his daughter's house, which he built alongside his own. He explains his resistance through a profound relationship with the natural environment where danger is neutralized by respectful coexistence: "Faults are everywhere—some bigger, some slower—but they're always there. When the lady, who I call my godmother [Etna], gets upset or has a stomach ache, that's when the slower ones … everyone gets scared! Not me, though, because I'm like an illegal settler on her land, but I sleep well, and so far, she's left me here" (Interview with Mr. Alfio). By contrast, Mr. Salvo—like most eligible beneficiaries—embraced the paradox of recurring disasters by accepting state-sponsored relocation funding.

> At first, I wasn't really on board with the idea—at least not personally. But looking at the situation, knowing I had a property on an active fault that would probably reactivate in about thirty years (because those Etna faults always reactivate sooner or later), I changed my mind. [...] So, the SCRAE, the reconstruction authority, used this ordinance that, at first, all the earthquake survivors—especially those living right on the fault in this narrow zone—maybe because it was the first time it was applied, didn't look on very favorably. But then, actually, I have to say that I, along with many others, probably most of us, realized we could buy [a new property] by relocating within the same municipality—or, more precisely, actually the ordinance allowed relocation "within nine municipalities." (Interview with Mr. Salvo)

The economic incentive enabled residents to put technical-scientific knowledge into practice, which they increasingly use to interpret their environment. The experience of the earthquake—and the fear it imprinted on their body—heightened risk perception, which expert knowledge helps to contextualise, ultimately fostering acceptance of relocation. Ms. Carla's account offers a compelling illustration of this process.

> There's one thing I'm truly grateful for: That they made us leave and didn't let us rebuild. Because honestly, I just couldn't go back there ... Even before we knew it was going to be torn down, we kept thinking, "how are we ever going to live in this house again?" Every time we stepped inside, it felt heavy—especially in the bedroom—like we were right back in those awful moments. We even thought about fixing it up just to sell it, because I was scared. Really scared. (Interview with Ms. Carla)

A widespread negativity toward the sites of the damaged homes—where the earthquake was physically felt in most cases—emerged in all interviews, to varying degrees. Many participants expressed unease, discomfort, or even rejection toward their former residences. Emotional ties to the past and to the concept of home seemed muted, as if they were the price paid for having lived along the fault line for years. Mr. Roberto feels a profound attachment to his homeland but recognizes the severity of the risks involved and accepts that rebuilding there is no longer a feasible option.

> I'd say emotions really play a big part—at least for me, personally. It's not easy ... But when you think about it, from a health and safety point of view, it just doesn't make sense to put your life at risk. The emotional

> attachment is definitely there, though. I mean, the idea of staying in the same town, finding another place nearby, trying to imagine yourself in a new home, still in the same community ... It's something you hold on to. But in the end, we have to accept that those houses sitting right on the fault lines have to be left behind. [...] The pain you go through—everyone deals with it in their own way ... Some people just don't want to hear about it anymore, don't want to see anything. But me, I say: Okay, it's gone, but there's still something left, a reference point I can go back to and see how it used to be. (Interview with Mr. Roberto)

These words reveal the creation of an "experiential map" (Low 2003, Basso 1996) that guides orientation amid *spaesamento*—a state of disorientation—allowing him to "go and see" former places, and partly feel what living along the fault line was like. The tension between disorientation and re-grounding intensifies when return is impossible, due to fear or the need to move forward. Rather than enduring ties to place or community, many residents preserve personal continuity through emotional attachment to belongings and "meaningful objects," as reflected in Ms. Michela's narrative.

> Before they tore it down, I had emptied the whole house—or at least I thought I had. But while they were demolishing it, this little cup came out, like one of those nice ceramic wedding favors ... There was this mountain of rubble, and that thing was just sitting on top, like it had fallen there. I took it home with me. At home, I have a lot of broken things I couldn't bring myself to throw away. My kids will probably get rid of them one day—and fair enough, they're just broken things—but to me, they're mine, even if they're broken. I even kept things that were chipped. That was my life, and I held onto it.

The synecdochic identification of the house's physical reconstruction with the broader reconstitution of the family unit—often strained by the earthquake—underscores the emotional and social processes of *appaesamento* experienced by many displaced residents.

This dynamic influences both the quest for a new place that maintains or rekindles close kinship bonds and the resistance to relocation fuelled by the desire to stay rooted where intimate relationships once flourished. Mr. Sebastiano's account of his mother offers a compelling insight into these dynamics, illustrating how attachment to place and the spatial configuration of the home are profoundly intertwined within a broader framework of "social geography" of emotions (Atkinson & Jacobs, 2016).

FIGURE 6.2 One of the houses demolished and relocated from one of the most severely affected streets in the hamlet of Fleri
SOURCE: AUTHORS' PHOTOGRAPH, TAKEN DURING FIELDWORK.

> She's ninety-two years old ... For her, moving to a different house, with new neighbours and new friendships, was a real shock. The first year was hard—she felt like a stranger in her own home. Then, little by little, she started to feel a bit better, but she never really got over it. In fact, every now and then, when something's wrong, she says "If I were still in my old house, I'd have the support of my neighbours" ... They'd all been living together for more than forty years, raised their kids side by side—it was like mutual support. They'd exchange advice, maybe meet in the evening to talk or watch TV together. That's how it had always been. And now she says, "Tonight I'm alone at home. If I were still in the old house, I could have a chat with my neighbours, we'd plan something" ... Everything changed. People like me, for example, we go out during the day to work, to take care of things—so we move on more quickly. But not the elderly. For them, it was tough—like being uprooted, like moving to another town and having to start all over again. For older people, it was both a trauma and a tragedy.

As the preceding discussion suggests, the process of relocation can be interpreted in two distinct yet interconnected ways. On the one hand, it represents an adaptation to a geophysical hazard recognized not only by the SCRAE and risk management institutions, but also by the survivors themselves. On the other hand, it constitutes an ambivalent experience—one that may either exacerbate or mitigate pre-existing conditions of socioeconomic vulnerability, depending on how these are perceived and navigated by individuals. Such vulnerabilities take on differing significance across generational lines, in particular.

In some cases, among the 120 families affected by the Ordinance, the relocation process was not immediately accepted. In "defending" the house built by his father, where he lived with his parents, Mr. Antonio initially expressed strong resistance to the safety zone established by the reconstruction authority—specifically, the geological map that, based on fault tracing, delineates areas where reconstruction is prohibited. Notably, SCRAE was the first institution in Italy to assume the responsibility of precisely defining such a zone—thus exposing itself to potential legal disputes—in order to safeguard the lives and property of those residing in areas of high seismic hazard. The following excerpt illustrates how he experienced his relationship with the SCRAE.

> I went several times to talk with the deputy commissioner; he was never convinced about this [my idea of rebuilding the house where it was], not because he wanted to stop me from building, but he always believed that [there] was a fault. We did some investigations that showed the fault and also disturbances shifted from the fault. We had the chance [to rebuild on the same land but moving about twenty meters away], but once again, the commissarial structure denied us permission to build. Then we got the opportunity to do a proper seismic analysis, and now I don't remember the names [of the geologists involved] … And things moved forward. So, it wasn't a fight between me and the reconstruction authority; for me, it was a defeat: not being able to rebuild [in the same place or nearby]. (Interview with Mr. Antonio)

This exchange reveals a process of assessment and negotiation involving the mobilization of technical expertise by both the individual concerned and SCRAE. Mr. Antonio, in particular, consulted independent professionals who, together with researchers specialized in paleoseismology, carried out a stratigraphic trench investigation. This analysis confirmed the kinematic activity of the fault at depth: The surface deformations were indeed caused by a distal termination of the fault, as initially indicated by the SCRAE to justify the mandatory relocation.

## 5 The Faul Line Agreement

The relationship between earthquake survivors and the SCRAE developed through a series of ongoing meetings and consultations held in the months following the event, led directly by the commissioner and deputy commissioner, both in the field and at SCRAE offices. It was precisely this regularity and the individualized technical dialogue with each family that helped foster a climate of relative trust.

Conversely, a significant mistrust towards the municipal administration emerged throughout the interviews, with survivors accusing it of generating delays, inertia, and a lack of transparency in procedures. The following are the grievances voiced by Mr. Antonio concerning the Autonomous Accommodation Allowance (CAS):[7] "The CAS is always late. Not because of the Commissioner or the government, but—as I see it—it's because the Municipality has called us many times, and I understand what the problem is: They're always behind, not just by a month, but by five or six months! So, I have to pay rent myself" (Interview with Mr. Antonio).

Field interviews reveal the challenges municipal offices face in managing funds from the national civil protection agency. Beyond delays in CAS disbursement, many also report slow progress in securing the road infrastructure.

This fosters residents' sense of neglect—both practical and aesthetic—undermining reconstruction progress and raising concerns about its overall pace, as Mr. Tommaso explains in the following excerpt: "You say, 'What's going on with this earthquake? Isn't it changing?' But that should be a sign—a sign of recovery, right? Like, 'Things are starting to move here.' Instead, this was something that got a bit overlooked" (Interview with Mr. Tommaso).

However, one of the main tasks the committees took on was gathering and sharing information with earthquake survivors specifically about the delays in the CAS.

These committees emerged spontaneously in the aftermath of the earthquake, as recounted by Mr. Sebastiano, an active member.

7 It is an economic assistance measure introduced in Italy following the 2009 L'Aquila earthquake to provide temporary support to people affected by seismic events in the immediate aftermath. CAS consists of a monthly financial contribution granted to families who cannot live in their damaged homes, allowing them to independently rent accommodation on the private market.

FIGURE 6.3 Road partially collapsed due to the earthquake, not yet repaired by local authorities
SOURCE: AUTHORS' PHOTOGRAPH, TAKEN DURING FIELDWORK.

FIGURE 6.4
Banner with the message "Fleri Rise Up. We will rise from the ashes", affixed to the façade of the church in the hamlet of Fleri by local committees
SOURCE: AUTHORS' PHOTOGRAPH, TAKEN DURING FIELDWORK.

> Right after the earthquake, we set up local committees in every town—sometimes more than one—each focused on a specific area like a neighborhood or village. Then we created a single Etna-wide committee to coordinate everything, which was initially led by the person who later became Commissioner. We pushed for a survivor to be appointed Commissioner—someone who truly knew the situation. He was a former judge, and we told him: "You've been through this—you understand the problems. If we support your appointment, that's already half the battle." Usually, Commissioners arrive with no local knowledge, just the law. Thanks to this strategy and our collaboration with both the Commissioner and the Deputy Commissioner, we managed to get the Ordinance changed on several points, including the introduction of a demolition grant, which hadn't been included at first. (Interview with Mr. Sebastiano)

The people we engaged with in the field showed a certain deference toward the SCRAE—not solely due to the formal authority conferred upon it by the state. Their attitude likewise conveyed a value judgment regarding local engagement, prior institutional roles, and the effectiveness demonstrated by some officials in managing the relocation process. This judgment challenged the dominant narrative of power as purely repressive and unilateral in its imposition on earthquake survivors. In contrast, the committees emerge as authentic mediators guiding affected individuals through complex bureaucratic processes, as exemplified by Mr. Mirko's experience.

> The committee really played its part during the various meetings we had—they brought us face to face with the Commissioner many times, as well as with the local authorities here. All the communications about the different measures taken by the Commissioner came to me through the committee. Sometimes they matched what my technician already knew—or he even found out first—but other times, it was the committee that informed me. So I have to say, from that point of view, the committees were really useful.

Findings from the interviews suggest that the institutional approach of selective and assisted relocation opened a space for testing risk prevention models driven by a logic of "mutual convenience." Indeed, the socio-economic benefits of relocation positioned convenience as a key driver, mediator, and moderator of safety. As Oka (2021) observes, convenience is a fluid, multidimensional concept—linked not only to efficiency and effectiveness, but also to ease, practicality, and culturally shaped understandings of advantage. Among the families we engaged with, initial hesitation or resistance often gave way to a broadly shared attitude of pragmatic convenience, as the following comment clearly illustrates.

> It's about understanding, really—on a technical level—that it just makes more sense, like I was saying earlier. At first, I was thinking: it's just a matter of cost! I thought, fine, I'll pour a new slab, build a stronger structure so it can kind of float on the earthquake, you know? Stay right where I was. That's how I saw it. But in the end, it didn't make sense—not for the State, and not for me either. (Interview with Mr. Salvo)

In this context, the notion of convenience extends beyond managing the financial and temporal burdens of (re)construction. It involves strategically balancing safety and cost-efficiency, guided by a form of "technical rationality." As families and individuals navigate risks, they also gain socioeconomic, technical, and emotional benefits from relocation. The issue is not merely maximizing utility in acquiring a new home through institutional assistance—a good limited by time, effort, and energy—but assessing how such investment effectively reduces risk. It is precisely around risk reduction that individual and state interests align, creating mutual benefit.

Perceiving a degree of technical control over the relocation process—as clearly expressed by Mr. Salvo and echoed by others—helps cast the effort toward housing safety in a positive light. Adapting to a more tangible post-earthquake risk landscape involves more than merely minimizing costs. Importantly, economic incentives (such as funding to purchase homes outside high-risk zones) proved effective only when paired with prompt action: Relocated households were prioritized by SCRAE, making rapid disbursement essential. Interviewees frequently emphasized the urgency of swiftly restoring their lives, often favoring timely relocation over rebuilding "as and where it was." As Mr. Mirko's account illustrates, it is the timing of recovery that ultimately defines the real dividing line across the affected territory: "I said to myself: I have to try to solve this problem, and I have to do it using the tools the State gives me—but as fast as possible. After four years and four days, we had a home again. But it was a different kind of home, because we didn't rebuild" (Interview with Mr. Mirko).

Several interviewees highlighted delays in in-situ reconstruction, which is subject to a procedural framework distinct from that of relocation. The following excerpt captures Mr. Roberto's assessment of this issue.

> I put myself in the shoes of those hundreds of families [earthquake survivors whose homes weren't exactly on the fault line] who still don't have a house. Hardly any homes have been built—so where is the State? It doesn't exist; the State is just something people say ... I don't really see it as a force that helps people in these situations. I'm not talking about myself—in one way or another, I managed to have a roof over my head. (Interview with Mr. Roberto)

The sense of the State's absence is also evident in the statements of those who had to restart their businesses. Below, Mr. Luigi's words capture the views shared by several merchants we encountered in the field.

> One thing I blame the State for is not providing any support to businesses. In other earthquakes, they gave grants and one-time payments for lost revenue in addition to property aid. Because of this, I've spent five or six years unable to pay anything, accumulating overdue taxes and debts to the Revenue Agency. I had to arrange payment plans because the State abandoned us. And because of that, I still haven't renovated the other property I bought for the relocated shop—I can't settle in and keep paying rent here. (Interview with Mr. Luigi)

Like the fault line itself, the State acts selectively—and is selectively perceived: By some, it simply "does not exist," except as a tax-collecting entity; for others, it seeks grounds of mutual convenience. This logic of convenience aligns more closely with the idea of an agreement than with that of imposed authority—an agreement negotiated on fragile terrain, where the relocated citizen and the commissarial body meet: The fault line itself. Precisely because it is selective, this "fault-line agreement" fosters collaboration between SCRAE and citizens, either individually or collectively through committees. In this context, the State operates as an influential agent that—albeit selectively—protects human life. It also serves as a technical–administrative interface, acting as driver, mediator, and moderator of safety within a framework of mutual benefit. Still, the uneven pace of reconstruction—shaped by divergent procedures, distinct authorities, and the varied purposes of affected buildings—complicates and renders public perceptions of institutions more ambivalent within their multi-level territorial articulation. The national government is absent in supporting the recovery of local businesses—perceived as too distant and "politically difficult to access," in the words of a prominent committee representative—while municipalities neglect territorial care and delay the use of state funds, yet adopt a legalist approach toward earthquake survivors, strictly enforcing regulations. In the following statement, Mr. Mirko hints at emerging criteria for assessing the "convenience" of relocation.

> We tried to rebuild on site—either by moving to another part of our large plot or looking for land nearby—because at first you get that feeling, like an instinct for self-preservation, for safety. You feel protected in your own place. But then you start to ask yourself, "At what cost? Do I really feel safe here? Is it worth it?" Building here would cost me ten times more than normal, because I'd need to use anti-seismic systems, shock absorbers, and all kinds of things, and I'd have to accept compromises. So is it really worth it? And then you look around at the state of the area, how it's been maintained and equipped. Because I think the context really influences you—if you're in a town, in a region that's well equipped, with public transport, schools, shops … but we didn't have any of that anymore here.

These words reveal how the principle of relocation as a safety strategy has increasingly supplanted place attachment, privileging future security over historical continuity. The process of *appaesamento*—and, by extension, the perceived legitimacy of the fault line agreement—unfolds through an ongoing assessment of everyday spatial advantages and a continual adaptation to the evolving landscape. This transformation is shaped by a reconstruction process marked by uneven temporal and spatial disparities.

## 6 Conclusions

Geophysical phenomena on the slopes of Mount Etna—particularly its eastern flank—are a persistent source of fear for local residents. As noted at the outset, seismic and tectonic activity will persist as long as the fault remains active. Earthquakes and lateral eruptions are deeply embedded in the existential landscape of those living in the area. Yet risk perception affects the "intimate spaces" of everyday life in varied ways (Ciccozzi 2016), shaping diverse personal and collective responses. The effort to ensure public safety—and, by extension, to promote environmentally grounded forms of life—must engage with the fluid and evolving boundaries of the body–territory relationship. Investigating the dynamics of *spaesamento* and *appaesamento* following the 2018 Etnean disaster involves examining the politics and practices of dwelling, highlighting how ill-conceived, uncoordinated, or rushed technical decisions can, over time, inflict harm greater than that of the initial event itself.

Exposure to potential harm—what disaster science typically refers to as "hazard"—cannot be fully captured by geophysical maps, or at least not in a neutral or objective way. Social representations of risk generate layered experiential and affective stratigraphies that resist linear cause-effect interpretations, especially when disasters enter a latent phase, embedding themselves in the folds of everyday life where their occurrence—real or potential—often escapes full awareness. Risk situations are always tied to the "strategies and mechanisms used to calm, confront, and overcome fear" (Boscoboinik & Horakova 2014: 10), which are inevitably shaped by cultural and social contexts. In this sense, an anthropological perspective is really essential. In contexts like these, the interplay between risk perception, anxiety management, and fear enables prudent, experience-based responses. Legitimate fear may even be a "privilege" (Humphrey 2013: 13), a "positive strategy that can make the difference between dying and staying alive" (ibid.). Yet, when threats remain invisible or only partially perceptible—surfacing through minor, fragmented incidents over time—preparation and prevention become more difficult, especially before the paroxysmal onset of disaster.

The relevance of this case lies in the SCRAE's effort to initiate a consultation process that did not automatically cast citizens as lay people incapable of grasping territorial risk or taking informed decisions. The chance to engage technically with the reconstruction authority enabled survivors to address legitimate doubts about the safety of their homes and the advisability of relocation—thus reclaiming agency and self-determination. The selective relocation approach adopted by the reconstruction authority stands not merely as an administrative or technical intervention, but as a dynamic social laboratory where notions of vulnerability, rootedness, and agency are continuously reconfigured in the aftermath of disaster. Rather than a straightforward top-down mandate, the ordinance application unfolded as an interactive field where institutional knowledge intersected with lived experience and place-based understandings, revealing underlying tensions and ambiguities. While relocation disrupted pre-existing social ties and collective routines, it simultaneously opened new avenues for individuals to reassert control by engaging with scientific frameworks and seizing spatial and material opportunities. Technical and scientific knowledge functions as both a cognitive and practical tool, helping to counteract the fear triggered by the event and to restore a risk perception that had been dulled by prolonged coexistence with danger.

While older residents often find it difficult to sever the web of relationships they have long inhabited, most of the relocated individuals we interviewed express satisfaction in providing their children with better homes—homes situated in areas offering greater comfort and improved services. Acceptance of relocation did not signify mere passive compliance or outright resistance; rather, it emerged as a context-specific adaptive strategy shaped by history, collective memory, and aspirations for a safer future. This case thus challenges simplistic binaries between state expertise and community agency, instead depicting disaster response as a negotiated arena where governance, knowledge production, and everyday life intersect. It underscores the necessity of policy approaches that recognize this complexity and foster inclusive processes—particularly when contrasting the perspectives of a commissioner's office directly engaged with the territory and capable of dialogue, versus an absent or overly bureaucratic state presence. Such approaches must be sensitive both to structural constraints and to the situated, lived ways people engage with risk and uncertainty.

## Bibliography


Atkinson, R., Jacobs, K., 2016. *House, Home and Society*, Palgrave, London.

Azzaro, R., Pucci, S., Villani, F., Civico, R., Branca *et al.*, 2022. "Surface Faulting of the 26 December 2018, Mw5 Earthquake at Mt. Etna Volcano (Italy): Geological Source Model and Implications for the Seismic Potential of the Fiandaca Fault." *Tectonics*, 41 (12). DOI: https://doi.org/10.1029/2021TC007182.

Bankoff, G., 2003. *Culture of Disaster: Society and Natural Hazards in the Philippines*, Routledge Curzon, London.

Barrios, R.E., 2011. "If You Did Not Grow Up Here, You Cannot Understand Living Here: Neoliberalism, Space-Time, and Affect in Post-Katrina New Orleans." *Human Organization* 70(2): 118–127.

Barrios, R.E., 2017. *Governing Affect: Neoliberalism and Disaster Reconstruction*. University of Nebraska Press, Lincoln.

Barrios, R.E., 2017. "The Anthropology of Crises and Disasters at the Onset of the Anthropocene." *Annual Review of Anthropology* 46: 151–166.Basso, K.H., 1996. *Wisdom Sits in Places: Landscape and Language Among the Western Apache*, University of New Mexico Press, Albuquerque.

Benadusi, M., 2015. "Antropologia Dei Disastri. Ricerca, Attivismo, Applicazione." *Antropologia Pubblica* 1 (1–2), 25–46. DOI: https://doi.org/10.1473/anpub.v1i1-2.85.

Bolin, R.C., Stanford, L., 1999. *Constructing Vulnerability in the First World: The Northridge Earthquake in Southern California*, in A. Oliver-Smith, S.M. Hoffman (eds.), *The Angry Earth: Disaster in Anthropological Perspective*, Routledge, New York, pp. 89–112.

Boscoboinik, A., Horakova, H., 2014. *The Anthropology of Fear: Cultures Beyond Emotions*, LIT Verlag, Munster.

Button, G., 2010. *Disaster Culture: Knowledge and Uncertainty in the Wake of Human and Environmental Catastrophe*, Left Coast Press, Walnut Creek.

Ciccozzi, A., 2016. "Forms of truth in the trial against the Commission for Major Risks: Anthropological notes." *Archivio Antropologico Mediterraneo* 18 (2), 65–82. DOI: https://doi.org/10.7432/AAM180205.

Civico, R., Pucci, S., Nappi, R., Azzaro, R., Villani, F., Pantosti, D., Venuti, A., 2019. "Surface ruptures following the 26 December 2018, Mw 4.9, Mt. Etna earthquake, Sicily (Italy): EMERGEO Working Group (Etna 2018)." *Journal of Maps*, 15(2), 831–837. DOI: https://doi.org/10.1080/17445647.2019.1683476.

Comaroff, J., Comaroff, J., 2019. *Etnography and the Historical Imagination*, Westview Press, Boudler.

De Martino, E., 2015 (1948). *The World of Magic*, HAU Books, Chicago.

De Martino, E., 2023 (1977). *The End of the World: Cultural Apocalypse and Transcendence*, Seagull Books, Chicago.

Diamond, J., 2005. *Collapse: How Societies Choose to Fail or Succeed*. Viking Press, New York.

Douglas, M., 1986. *Risk Acceptability According to the Social Sciences*. Russell Sage Foundation, New York.

Douglas, M., 1992. *Risk and Blame Essays in Cultural Theory*. Routledge, New York.

Douglas, M., Wildavsky, A., 1983. *Risk and Culture: An Essay on the Selection of Technological and Environmental Dangers*, University of California Press, Berkley and Los Angeles.

Falconieri, I., Fichera, F., Valitutto, S., 2021. *Irpinia 1980: evocare il terremoto, ripensare i disastri*. C&P Adver Effigi, Grosseto.

Gugg, G., 2013. *All'ombra del Vulcano. Antropologia del rischio di un paese vesuviano*, PhD, University of Naples L'Orientale.

Gugg, G., 2017. "Rischio e post-sviluppo vesuviano." *Antropologia Pubblica*, 1 (1–2), pp. 109–124. DOI: http://dx.doi.org/10.1473/anpub.v1i1-2.91.

Gunewardena, N., Schuller, M. (eds.), 2008. *Capitalizing on Catastrophe: Neoliberal Strategies in Disaster Reconstruction, (Globalization and the Environment*), AltaMira Press, Lanham.

Humphrey, C., 2013. "Fear as a Property and an Entitlement." *Social Anthropology*, 21 (3), pp. 285–304. DOI: https://doi.org/10.1111/1469-8676.12033.

Kartol, A., Üztemur, S., Yaşar, P., 2024. "Development and Validation of the Earthquake Obsession Scale." *Death Studies*, 1–9. DOI: https://doi.org/10.1080/07481187.2024.2317177.

Klein, N., 2007. *The Shock Doctrine: The Rise of Disaster Capitalism.* Knopf Canada, Toronto.

Low, S.M., Lawrence-Zúñiga, D. (eds.), 2003. *The Anthropology of Space and Place: Locating Culture.* Blackwell, Oxford.

Lupton, D., 1999. *Risk.* Routledge, London.

Lupton, D., 2013. "Risk and Emotion: Towards an Alternative Theoretical Perspective." Health, Risk & Society 15 (8): 634–47. DOI: https://doi.org/10.1080/13698575.2013.848847.

Masco, J., 2004. "Mutant Ecologies: Radioactive Life in Post–Cold War New Mexico." Cultural Anthropology 19 (4): 517–550. DOI: https://doi.org/10.1525/can.2004.19.4.517.

Mattia, M., Bruno, V., Caltabiano, T., Cannata, A. *et al.*, 2015. "A Comprehensive Interpretive Model of Slow Slip Events on Mt. Etna's Eastern Flank." *Geochemistry, Geophysics, Geosystems*, 16 (3), pp. 635–658. DOI: https://doi. org/10.1002/2014GC005585.

Moscaritolo, G.I., 2020. *Memorie dal cratere. Storia sociale del terremoto in Irpinia*, Editpress, Florence.

Neri, M., Neri, E., 2025. "Geoethics and the 2018 Post-Earthquake Recovery on Etna: Psychological Consequences of Relocation Strategies". *Preprints.* DOI: https://doi.org/10.20944/preprints202505.2283.v1.

Oka, R., 2021. "Introducing an Anthropology of Convenience." *Economic Anthropology*, 8 (2), pp. 188–207. DOI: https://doi.org/10.1002/sea2.12219.

Oliver-Smith, A., 1999. *"What Is a Disaster?" Anthropological Perspectives on a Persistent Question*, in A. Oliver-Smith, S. Hoffman (eds.), *The Angry Earth: Disaster in Anthropological Perspective*, Routledge, London and New York, pp. 158–172.

Oliver-Smith, A., 2010. *Defying Displacement: Grassroots Resistance and the Critique of Development*, University of Texas Press, Austin.

Oliver-Smith, A., Hoffman, S. (eds.), 1999. *The Angry Earth: Disaster in Anthropological Perspective*, Routledge, London and New York.

Petino, G., Napoli, M.D., Mattia, M., 2023. "The Belìce Valley as a Territorial Laboratory: from Public Policy Experimentation to a Large-area Administration Strategy." *Semestrale di studi e ricerche di geografia*, (2), pp. 129–152. DOI: https://doi.org/10.13133/2784-9643/18314.

Satici, A.S., Sinan, O., Deniz M.E., Karaağaç, Z.G. *et al.*, 2023. "The Development and Initial Validation of the Earthquake Fear Scale: Its Links to Personality Traits, Psychological Distress, Larmony in Life, and Mental Wellbeing." *Stress & Health*, 40 (2), e3306. DOI: https://doi.org/10.1002/smi.3306.

Signorelli, A., 1992. "Cathastrophes naturelles et résponses culturelles." *Terrain. Anthropologie & sciences humaines*, 19, pp. 147–158. DOI: https://doi.org/10.4000/terrain.3052.

Simpson, E., 2020. "Forgetfulness Without Memory: Reconstruction, Landscape, and the Politics of the Everyday in Post-Earthquake Gujarat, India." *Journal of the Royal Anthropological Institute*, 26 (4), pp. 786–804. DOI: https://doi.org/10.1111/1467-9655.13416.

Ullberg, B.S., 2010. "Disaster Memoryscapes: How Social Relations Shape Community Remembering of Catastrophe." *Anthropology News*, 51 (7), pp. 12–15. DOI: https://doi.org/10.1111/j.1556-3502.2010.5.

CHAPTER 7

# The Role of the Disaster Museum in Japan: Inheriting Memory and Promoting Disaster Education

*Bonjun Koo*

## 1 Introduction

Japan's experience with natural disasters is unparalleled in both frequency and scale. The archipelago's geographic location at the intersection of multiple tectonic plates—namely, the Pacific, Philippine Sea, Eurasian, and North American plates—renders it uniquely vulnerable to earthquakes, tsunamis, and volcanic eruptions (Bird, 2003). Additionally, its exposure to seasonal typhoons, heavy rainfall, landslides, and occasional cold-weather events has produced a multihazard landscape that demands a culture of constant preparedness. These recurring shocks are not just natural occurrences—they are deeply inscribed into the nation's psyche, architecture, and public policy. From the earliest recorded disasters in historical chronicles to the technologically sophisticated alerts of the modern era, disaster awareness is a defining feature of Japanese civic life.

Throughout history, major catastrophes have profoundly shaped Japanese governance, education, and spatial planning. Historical records, such as those from the *Nihon Shoki* or the *Azuma Kagami*, recount floods, earthquakes, and famines that decimated provinces and compelled institutional change (Nihon Shoki, 1994; Azuma Kagami, 1997). These events cultivated a collective consciousness that regards disasters not as exceptional disruptions but as recurring episodes that must be anticipated. Over generations, this worldview has nurtured a broad-based culture of vigilance, ritualized preparedness, and communal responsibility. Even in popular culture, from Edo-era woodblock prints to modern anime, disaster motifs convey moral and political lessons about impermanence, adaptability, and social cohesion.

In response, Japan has developed one of the world's most advanced disaster risk management systems. This system includes state-of-the-art seismographic and meteorological monitoring, an extensive network of early warning systems, and legal frameworks that mandate disaster education and preparedness drills across society (Cabinet Office, 2022; Ministry of Internal Affairs and

 | HTTPS://DOI.ORG/10.1163/9789004751804-009

Communications, 2021). However, beyond the institutional apparatus, Japan's emphasis on community-based approaches distinguishes its resilience model. Voluntary disaster preparedness organizations (*bōsai kurabu*), neighborhood councils, school-based evacuation protocols, and resident-led hazard mapping are hallmarks of this decentralized, socially grounded approach.

The 1995 Great Hanshin-Awaji Earthquake marked a watershed moment in modern Japanese disaster governance (Hyogo Prefecture Reconstruction Headquarters, 1996). It revealed the limitations of top-down emergency response and highlighted the vital role of local actors and civil society in recovery efforts. In its aftermath, the Disaster Reduction and Human Renovation Institution (DRI) was established as a center for documentation, education, and innovation in disaster resilience. Located in Kobe, one of the cities most affected by the quake, the DRI embodies a new paradigm that integrates technical expertise with public memory, emotional healing, and international exchange.

Moreover, the Great East Japan Earthquake of 2011 further transformed Japan's understanding of disaster governance (UNDRR, 2015; Cabinet Office, 2012). The compounded crisis of an earthquake, tsunami, and nuclear accident demonstrated the need for more integrated, flexible, and inclusive risk management systems. In particular, it emphasized the importance of non-structural measures such as psychosocial support, community empowerment, and transparent risk communication. These lessons have informed not only national policy reforms but also global frameworks like the Sendai Framework for Disaster Risk Reduction (2015–2030), which Japan helped shape.

The evolving discourse around resilience in Japan has increasingly embraced holistic principles—linking physical safety with social cohesion, infrastructure with emotion, and science with culture. Resilience is no longer understood as merely "bouncing back" from disasters; it is being reconceptualized as "building forward" with inclusivity, justice, and sustainability at its core. Schools, neighborhoods, research institutes, and city halls have become active sites of collaborative preparedness, where disaster mitigation is understood as a process of civic empowerment. Furthermore, resilience has become a touchstone for addressing intersecting crises—including demographic aging, climate change, and urban inequality—each of which amplifies disaster risk in distinct ways.

Academic institutions and think tanks now contribute to resilience discourse not only through technical research but also through interdisciplinary collaborations that include the arts, humanities, and social sciences. Citizen science initiatives, for instance, enable residents to participate in environmental monitoring and historical documentation of local hazards. Universities run fieldwork programs in disaster-affected communities, where students learn firsthand about memory, ethics, and planning under uncertainty. These engagements exemplify Japan's deepening investment in resilience as a socially and culturally informed practice.

This chapter introduces a multifaceted exploration of Japan's approach to disaster and recovery, with particular attention to how community structures, memory institutions, and inclusive planning have enabled adaptive responses to past and ongoing hazards. It integrates historical perspectives with contemporary policy developments and case studies to show how resilience is not a static outcome but a continually negotiated process. Furthermore, by situating Japan within the global context of disaster risk reduction—especially through its leadership in formulating international frameworks such as the Hyogo Framework for Action and the Sendai Framework—the chapter also reflects on how national experience informs and is informed by global discourse.

The subsequent sections unpack the evolution of Japan's community-based disaster practices, the institutional structure and pedagogical philosophy of the DRI, and several critical case studies illustrating the complexity of post-disaster recovery. Through this narrative, the chapter foregrounds the human and relational dimensions of resilience and advocates for a holistic approach to disaster governance that prioritizes empathy, memory, and collective imagination. It also bridges academic analysis with practitioner insights and community narratives, underscoring that the task of building resilience is both technical and moral, both historical and visionary.

## 2 Historical Foundations of Community-Based Disaster Responses

Long before the establishment of modern disaster management systems, Japanese communities relied on deeply rooted practices of mutual aid, local knowledge, and religious guidance to cope with natural hazards. These historical foundations continue to inform Japan's contemporary community-based disaster risk reduction (CBDRR) strategies.

During the Edo period (1603–1868), agrarian communities practiced *yui* (mutual assistance) and *moyai* (collective work), systems of reciprocal labor for agricultural tasks, house building, and post-disaster recovery. These practices were not codified in law but were governed by social norms, oral agreements, and the embedded expectation of community obligation (Aldrich, 2012). Such customs reinforced group cohesion and ensured that no household was left unaided in the wake of floods, fires, or earthquakes. Beyond their functional value, these systems reflected the philosophical underpinnings of Japanese society, such as Confucian ethics of duty and Buddhist teachings on impermanence and interconnectedness (Reader, 1991).

Temples and shrines also functioned as proto-institutions of disaster management. In addition to providing spiritual solace, they acted as physical refuges, coordinated donations of food and supplies, and maintained records of local hazards and events (Nelson, 2003). Buddhist cosmology and Shinto rituals, which emphasized purification, cyclical renewal, and reverence for nature, helped frame disaster experiences within a culturally intelligible worldview. Annual rites and festivals often commemorated past disasters, turning memory into ritualized resilience. Local deities were sometimes venerated for their protective roles against fire or flood, and spiritual leaders assumed practical responsibilities in organizing community-based response. In some regions, specific temples became known as sanctuaries during floods or earthquakes and developed permanent stockpiles of emergency goods.

With the Meiji Restoration (1868), Japan embarked on a process of modernization and centralization. Western science, engineering, and governance were imported to enhance the state's ability to manage risk. New infrastructures—railways, river embankments, fire departments—coexisted with traditional communal practices. The 1923 Great Kanto Earthquake, which devastated Tokyo and Yokohama, marked a turning point (Clancey, 2006). The collapse of state authority in the disaster's immediate aftermath elevated the role of neighborhood associations (*chōnaikai*) and volunteer fire corps (*shōbōdan*). These groups coordinated evacuations, built temporary shelters, and distributed resources. Over time, they became semi-formal civic institutions, later supported by legislation such as the 1947 Disaster Relief Act and the 1961 Disaster Countermeasures Basic Act. These legal frameworks laid the groundwork for a hybrid system of top-down and bottom-up risk governance.

In the postwar period, Japan's rapid urbanization and economic growth transformed its social and spatial fabric. Yet, community-based responses remained salient. During the 1964 Niigata Earthquake and the 1978 Miyagi Earthquake, newspapers and government reports praised local efforts to rescue trapped residents and maintain order before outside help arrived (Aldrich, 2011). Academic research by disaster sociologists in the 1980s and 1990s increasingly emphasized the importance of local autonomy, trust networks, and informal communication channels in effective disaster response. The notion of "self-help" (*jijo*), alongside "mutual help" (*kyōjo*) and "public help" (*kōjo*), became central to national disaster policy frameworks (Shaw & Goda, 2004).

Today, *chōnaikai* remain foundational to Japan's community preparedness. They are responsible for organizing annual disaster drills, disseminating emergency warnings, and maintaining registries of vulnerable residents, such as the elderly and people with disabilities. In many neighborhoods, they oversee the distribution of disaster kits, training of volunteer brigades, and coordination with local schools and hospitals. These functions are often carried out in collaboration with municipal governments and are supported by subsidies, technical manuals, and training from prefectural agencies. The resilience of these groups was particularly evident during the 2004 Niigata Chuetsu Earthquake and the 2016 Kumamoto Earthquakes, where early rescue and evacuation actions were led by community members themselves (Shaw, 2014).

Contemporary CBDRR initiatives have also integrated global concepts such as "participatory mapping," "inclusive design," and "social capital enhancement." Programs facilitated by NGOs and universities encourage communities to produce "hazard maps" that identify evacuation routes, safe spaces, and high-risk zones (Shaw, 2014). Intergenerational workshops, storytelling sessions, and mock evacuation games foster a culture of preparedness, particularly among children and youth. Additionally, community radio networks, digital bulletin boards, and multilingual emergency apps have emerged as tools for bridging communication gaps in multicultural neighborhoods. These innovations demonstrate how traditional forms of mutual support can be reconfigured through modern tools, especially in increasingly diverse urban contexts.

The evolution of community-based disaster responses in Japan illustrates a synthesis of past and present, local and global. Traditional practices such as *yui* have found new expression in volunteer networks. Religious rituals of mourning now intersect with psychological first aid and post-traumatic care. And grassroots institutions like the *chōnaikai* have adapted to urban diversity, technological change, and the growing complexity of compound disasters. Recent government initiatives have also introduced CBDRR guidelines into the formal school curriculum, reinforcing community-based knowledge at an early age (Cabinet Office, 2015).

This historical continuity is not static—it evolves through lived experience, reinterpretation, and institutional support. It underscores that resilience is not imposed from above, but cultivated from within, grounded in a legacy of collective responsibility and moral imagination. Japan's case suggests that a resilient society is not merely one that builds higher walls or stronger infrastructure, but one that continually invests in social cohesion, memory, and mutual care as core strategies of risk governance.

## 3 The DRI as a Memory Institution and Educational Platform

The Disaster Reduction and Human Renovation Institution (DRI), established in 2002 in Kobe by the Hyogo Prefectural Government with support from the Japanese central government, was conceived as a permanent institution to preserve the collective memory of the Great Hanshin-Awaji Earthquake of 1995 and to disseminate its lessons to future generations. Unlike conventional disaster museums, DRI represents a fusion of memorialization, public education, and professional training. It functions simultaneously as a museum, an international knowledge hub, a research center, and a space for emotional and community recovery. Its multifaceted mission encapsulates Japan's belief that remembering past disasters is central to building future resilience (DRI, n.d.-a).

The architectural layout of DRI reflects its dual function as a space of mourning and forward-looking preparedness. The West Building contains immersive exhibits, most notably the "1.17 Theater," which recreates the experience of the 1995 earthquake through synchronized audiovisual effects, personal testimonies, and real-time recordings. Visitors not only see the physical destruction but also encounter survivors' stories of trauma, rescue, and solidarity. The East Building focuses on reconstruction, policy learning, and global cooperation. It includes displays on long-term recovery, regional revitalization, and technological innovations in disaster management. Exhibits are multilingual and accessible, using visual, tactile, and narrative methods to engage diverse audiences—including school children, international researchers, and elderly residents (DRI, n.d.-b).

DRI's design philosophy draws from both Japanese aesthetics of impermanence and contemporary museology. The facility avoids overwhelming textual panels, instead inviting experiential learning through sensory engagement and guided narration. The use of wood, natural light, and open spaces is intentional, facilitating reflection and psychological safety. Its curatorial choices reflect a balance between honoring grief and fostering proactive learning. This spatial and pedagogical environment is particularly conducive to trauma-informed education and long-term memory work.

One of DRI's distinctive features is its emphasis on emotional recovery and the ethics of memory. Central to this is the Listening Archive (*Kikikomi Shiryō*), a curated oral history project that collects, preserves, and presents the voices of disaster survivors, first responders, and volunteers. These stories are not merely recorded—they are actively integrated into educational curricula, recovery planning workshops, and international exchange programs. For example, student visitors may watch survivor testimonies before participating in discussion groups or simulation exercises, cultivating empathy and critical reflection on the human aspects of disaster.

DRI's educational programs extend beyond disaster drills. They include scenario-based planning workshops, inclusive risk communication training, and cross-generational dialogues on memory and preparedness. Participants range from local youth to international policymakers. Through partnerships with institutions such as JICA and UNDRR, DRI regularly hosts training programs for disaster managers from Southeast Asia, South America, and Africa (UNDRR, 2025). These exchanges promote South–South cooperation and enable adaptation of Japan's community-based resilience strategies to other contexts.

Moreover, DRI's archives and documentation center offer a rich repository of academic research, policy documents, personal letters, and multimedia materials related to the Hanshin-Awaji Earthquake and subsequent disasters. This repository not only supports research and policy evaluation, but also helps shape collective memory and societal narratives. The DRI collaborates with universities, civil society organizations, and the media to ensure these materials are widely accessible and ethically presented. It has also supported the digitization of archival content, expanding its accessibility to a global audience and contributing to disaster memory scholarship.

Beyond documentation, DRI plays a key role in organizing the annual International Recovery Forum, one of the most important global events focused on post-disaster recovery. The forum invites disaster professionals, scholars, community leaders, and UN officials to share best practices, policy innovations, and unresolved challenges. Recent forums have addressed topics such as cultural heritage preservation, gender-inclusive recovery, mental health, and climate adaptation. DRI's role as host underscores its status not only as a local institution, but as a globally recognized platform for transdisciplinary and cross-sectoral dialogue (DRI, n.d.-c).

In recent years, DRI has expanded its research agenda through action-oriented collaborations. It facilitates field projects in partnership with municipalities, focusing on themes such as inclusive shelter design, community resilience indicators, and emotional recovery in post-trauma environments. One notable project involved co-developing multilingual emergency protocols for foreign residents in Kobe, which were later adopted by other cities across Japan. These protocols were tested through live drills, involving community interpreters, nursing staff, and city officials (Hyogo International Association, n.d.).

DRI's campus itself acts as a living laboratory. Public programs invite residents of Kobe and visitors from around the world to engage with exhibits that are regularly updated to include recent disaster cases such as the 2016 Kumamoto Earthquake and 2018 Western Japan floods. DRI's "Learning Stations" facilitate peer-led exchanges between survivors and high school students, creating an intergenerational learning platform that goes beyond conventional classroom education. In these workshops, youth interview elders about their disaster experiences and co-create multimedia outputs—such as podcasts, timelines, and mini-documentaries.

The institution also serves as a model for evaluating the long-term effectiveness of disaster education. Follow-up studies conducted in collaboration with local schools track changes in disaster literacy, risk perception, and preparedness behavior among students. These metrics help refine pedagogical strategies and inform policy recommendations. The evidence gathered supports the view that immersive, narrative-based education has a stronger emotional and behavioral impact than conventional lecture-based instruction.

Through its unique integration of remembrance, education, innovation, and global exchange, DRI stands as a model of how disaster institutions can serve society not only by archiving the past, but by equipping communities with the cognitive, emotional, and institutional resources necessary to confront the uncertain risks of the future. As both a solemn space and a lively educational center, DRI transforms disaster memory into a living resource—a cornerstone for building inclusive, just, and sustainable resilience systems worldwide.

## 4 Case Studies of Earthquake Response and Community-Driven Recovery

The role of community resilience in disaster response becomes especially evident when examining Japan's major contemporary earthquakes. While the 1995 Great Hanshin-Awaji Earthquake and the 2011 Great East Japan Earthquake remain landmark examples of community-driven recovery, additional cases such as the 2016 Kumamoto Earthquake and the 2018 Western Japan floods further illustrate the evolution and diversification of grassroots resilience strategies. These examples reveal the capacity of ordinary citizens, civil society organizations, and local governments to improvise and innovate during crises (Shaw & Goda, 2004).

In the immediate aftermath of the Hanshin-Awaji Earthquake, which struck Kobe and surrounding areas in the early morning of January 17, 1995, local residents were the first to respond. With major roads blocked, communication infrastructure damaged, and emergency services overwhelmed, neighbors rescued one another from collapsed buildings, set up makeshift shelters in parks and schoolyards, and organized food distribution. These grassroots efforts were later formalized through the emergence of *jishubo* (voluntary disaster preparedness organizations), which continue to play an essential role in community disaster readiness. The earthquake also catalyzed the development of

Japan's nonprofit sector, as many citizen groups and volunteers formed what would become long-term organizations engaged in social welfare, advocacy, and disaster education (Shaw & Goda, 2004).

Importantly, the Kobe experience also prompted local governments to shift from rigid command-and-control models to more flexible, collaborative frameworks. Community centers, known as *kōminkan*, became hubs for post-disaster engagement, hosting legal aid consultations, grief support sessions, and neighborhood planning discussions. DRI's researchers played a role in analyzing these emerging structures, contributing policy guidance to local and national governments on the institutionalization of community resilience (Aldrich, 2012).

The Great East Japan Earthquake of March 11, 2011, presented an even more complex challenge: a massive earthquake, followed by a devastating tsunami and the Fukushima Daiichi nuclear disaster. Here, community-driven recovery efforts extended far beyond emergency response. Villages such as Otsuchi and Rikuzentakata were decimated, with entire neighborhoods swept away. Yet, in the face of loss, survivors gathered in temporary shelters to grieve together, recreate community networks, and plan reconstruction. In some areas, residents opposed top-down relocation plans and instead proposed their own vision for safer, higher-ground settlements.

Universities, NGOs, and planning experts collaborated with local stakeholders in participatory workshops to co-design town redevelopment strategies. These workshops were typically facilitated by interdisciplinary teams comprising urban planners, anthropologists, social workers, and local government officials. Methods included small group dialogues, community mapping exercises, and future scenario planning using visual materials such as maps, models, and storytelling prompts. Efforts were made to ensure inclusivity by inviting representatives of women's groups, senior citizens, youth organizations, and persons with disabilities. Outcomes extended beyond physical reconstruction plans to include revised land use policies, the identification of culturally significant landmarks for preservation, and the design of multipurpose public spaces that would serve both daily life and future emergencies. In several towns, these participatory processes culminated in official community charters that documented shared visions, priorities, and planning timelines—giving residents a tangible sense of ownership over recovery decisions. These inclusive processes redefined the role of affected residents from passive recipients to active agents in recovery.

Notable among these efforts was the "*Machizukuri*" movement—community-based town-making—that gave rise to locally authored rebuilding plans. In Minamisanriku, a devastated coastal town, survivors worked with architects and facilitators to envision new housing clusters, redesigned public spaces, and multi-purpose evacuation facilities that doubled as community centers. The plans were not always implemented fully due to bureaucratic constraints, but they demonstrated the potential of participatory governance in shaping post-disaster futures.

Since then, *Machizukuri* principles have been partially institutionalized through Japan's urban planning and disaster recovery systems. Several local governments have created dedicated "*Machizukuri* Councils" that serve as advisory bodies during reconstruction projects. These councils often include a wide array of stakeholders—residents, NGOs, academic experts, and local officials—and are empowered to influence zoning decisions, architectural design standards, and budget allocations. Furthermore, lessons from the *Machizukuri* movement have informed national guidelines on community engagement during recovery, including the Cabinet Office's policy documents on inclusive town planning (Cabinet Office, 2020). Elements of this model have also been adopted internationally in Japan's overseas development assistance programs, especially in post-disaster reconstruction efforts in Southeast Asia. In this way, the ethos of *Machizukuri* continues to evolve as both a locally grounded and globally resonant approach to building back better.

Although DRI primarily focuses on the Hanshin-Awaji Earthquake, it has contributed to disaster education and awareness more broadly through nationwide exhibitions and outreach. Its activities after the Great East Japan Earthquake have mainly included public messaging and knowledge dissemination, rather than direct field engagement.

In Sendai, which hosted the Third UN World Conference on Disaster Risk Reduction in 2015, the city's efforts to institutionalize community participation became a model for urban resilience. The city implemented neighborhood hazard mapping with school students, conducted citywide evacuation drills that involved private businesses, and created multilingual disaster information systems to serve its growing foreign population. DRI collaborated with Sendai City on post-conference knowledge transfer, including a "Sendai School of Recovery Leadership" that trained young professionals from both Japan and abroad. One participant, a civil servant from the Philippines, later launched a localized hazard mapping initiative in her home municipality modeled after Sendai's participatory techniques. Another Japanese graduate cited the program's emphasis on ethical leadership and cross-sector coordination as pivotal in shaping her career in municipal emergency planning. These testimonials reflect the School's broader impact: cultivating a new generation of disaster leaders capable of blending local knowledge with global best practices.

These case studies highlight several critical lessons. First, they reveal that the timeliness and effectiveness of response are often determined by pre-existing social networks. Second, they underscore the importance of locally grounded knowledge and culturally sensitive planning. Third, they show that recovery is not a technocratic process but a deeply human endeavor that must account for memory, trauma, and place. Japan's experience has helped popularize concepts such as "Build Back Better," which integrates sustainability, equity, and community values into reconstruction. Finally, they illustrate that empowering survivors as co-creators of recovery fosters not only physical rebuilding but emotional and civic regeneration.

Through initiatives like the DRI, these lessons are not only preserved but transformed into educational content, shared with international partners, and applied to new contexts. The institutionalization of community-driven recovery in Japan stands as a powerful example of how resilience must begin with people—empowered, informed, and connected to the places they call home.

## 5 Emotional Recovery and the Politics of Memory

Disaster recovery is never purely a technical or logistical endeavor—it is also deeply psychological and cultural. In Japan, emotional recovery is recognized as a critical aspect of long-term resilience, and efforts to address trauma, grief, and memory are increasingly integrated into disaster governance frameworks. Emotional recovery encompasses not only mental health services but also rituals, storytelling, artistic expression, and acts of public remembrance. These practices are vital for rebuilding a sense of belonging and continuity, especially in communities that have experienced profound loss.

In the wake of the 1995 Great Hanshin-Awaji Earthquake, many survivors reported feelings of abandonment, guilt, and identity dislocation. While the physical reconstruction of Kobe progressed steadily, the emotional scars endured. Responding to this need, community centers, faith-based groups, and municipal mental health teams developed innovative programs combining counseling with community art, festivals, and spiritual care. These initiatives laid the foundation for what is now referred to in Japan as *kokoro no kea*, or care for the heart—a phrase that encapsulates the nation's approach to trauma healing. These early practices served as pilot models that were later incorporated into formal emergency response protocols across Japan. For instance, mental health professionals are now routinely deployed alongside rescue teams, and municipalities are required to maintain databases of trained counselors and support volunteers.

The Disaster Reduction and Human Renovation Institution (DRI) has been a pioneer in institutionalizing emotional recovery. Its Listening Archive presents the voices of survivors not just as historical data, but as living testimony. These oral narratives are displayed in audio rooms, subtitled videos, and interactive exhibits, where visitors can hear survivors' words while viewing images of their neighborhoods before and after the disaster. Rather than offering closure, the project encourages ongoing engagement with memory. Workshops that accompany the Listening Archive invite schoolchildren and visitors to respond with letters, drawings, or their own stories, creating a multi-generational dialogue around loss and resilience. Some schools have incorporated these activities into their ethics or social studies curriculum, ensuring that disaster memory becomes a part of civic education.

DRI also supports annual memorial events such as the 1.17 Memorial Gathering, where citizens light thousands of candles to form the date "1.17" in public squares across Kobe. This ritual acts as both a communal expression of grief and a reaffirmation of collective hope. Similar practices were adapted in northeastern Japan after the 2011 disaster, including lantern floating ceremonies, temporary shrines at tsunami-hit sites, and public poetry readings. DRI has documented these rituals through photography and interviews, curating traveling exhibits that explore the aesthetic and emotional dimensions of disaster mourning across regions.

The politics of memory is not without tension. In the post-2011 recovery process, competing narratives emerged around who was responsible, what should be remembered, and whose voices counted. For example, some residents felt their experiences were marginalized in national media, while others opposed government decisions to rebuild or memorialize certain areas. Grassroots museums, citizen archives, and temporary exhibitions became platforms for contestation and community-driven memory work. These spaces challenged official narratives and elevated alternative voices—especially those of elderly residents, children, and displaced families. In particular, projects led by women's groups and disabled persons' associations brought attention to overlooked needs and redefined what counts as recovery experience.

In this context, DRI's approach to memory is notable for its pluralism. The institution does not impose a singular interpretation of events but curates diverse perspectives through partnerships with survivors, artists, educators, and social scientists. Memory is treated not as static heritage, but as a living, relational process. Exhibits and archives evolve over time to include recent disasters and new forms of storytelling, such as digital animations, augmented reality reconstructions, and survivor podcasts. The introduction of multilingual story stations has further broadened participation, allowing foreign residents to contribute their own experiences of disaster and recovery in Japan.

Ultimately, emotional recovery in Japan is not confined to individual therapy or commemoration. It is embedded in schools, neighborhoods, religious practices, and civic culture. It recognizes that resilience is not simply the absence of trauma, but the capacity to live meaningfully with it. Institutions like DRI play a crucial role in fostering this capacity, offering spaces where mourning, learning, and hope are interwoven into the broader narrative of recovery. As Japan faces future disasters, this model of integrating emotional and cultural dimensions into risk governance may prove as vital as physical rebuilding. It demonstrates that healing is not a finite goal, but a continuous act of remembrance, relationship, and renewal.

## 6 Digital Innovation and the Future of Risk Governance

Japan's commitment to disaster preparedness and response extends well beyond traditional practices. In recent decades, the country has emerged as a leader in integrating digital technology into disaster risk governance. Innovations in artificial intelligence (AI), geospatial analysis, satellite imaging, blockchain, and big data analytics are increasingly being applied across the disaster cycle—from early warning to recovery. This technological evolution, while offering new efficiencies, also presents ethical, social, and infrastructural challenges that institutions such as the Disaster Reduction and Human Renovation Institution (DRI) are actively working to address.

At the level of early warning, Japan operates one of the world's most sophisticated earthquake and tsunami alert systems, operated by the Japan Meteorological Agency (JMA) and supported by networks of seismographs and ocean-bottom sensors. Residents receive alerts via television, radio, smartphone apps, and public loudspeakers—sometimes seconds before the shaking starts. These systems have saved thousands of lives and are continuously refined with support from national and university research labs. In addition, DRI helps educate communities on how to respond effectively to digital warnings through interactive simulations and training modules. These include school-based earthquake drills enhanced with AR-based hazard simulations and community workshops where residents interact with tablet-based evacuation route tools.

Digital mapping and participatory GIS have also become essential to community-level preparedness. Local governments and residents co-create hazard maps using open-source platforms, layering topographical data with past disaster records, population density, and critical infrastructure locations. This democratization of data allows for more inclusive planning and fosters localized ownership of disaster mitigation strategies. In many cases, DRI serves as a repository and dissemination node for these materials, offering annotated case studies that compare digital risk mapping practices across municipalities. Some of these projects have even been incorporated into school curricula and senior center workshops as participatory education tools.

In the realm of post-disaster recovery, digital platforms have been developed to track resource distribution, manage volunteer logistics, and assess damages in real time. For example, during the 2016 Kumamoto Earthquakes, social media data was analyzed to identify areas in need of urgent support. Similarly, blockchain technologies have been piloted for transparent aid tracking, reducing the risk of corruption or duplication. DRI has facilitated workshops on these technologies, bringing together engineers, policymakers, and residents to explore their implications and limitations. In one such initiative, disaster apps were evaluated in multilingual user tests that involved foreign residents and visually impaired users, highlighting gaps in usability that informed subsequent design updates.

Yet, as digital tools expand, so too do concerns about access and equity. Elderly populations, individuals with disabilities, and residents in rural or underserved areas often face digital divides that limit their ability to benefit from new technologies. Moreover, questions around data privacy, surveillance, algorithmic bias, and cyber-resilience are increasingly central to disaster governance debates. DRI addresses these issues through public forums and policy roundtables that aim to bridge the gap between technologists and the communities they serve. The institution also produces policy briefs summarizing community feedback and collaborates with government agencies to revise digital inclusion strategies. In one pilot program, DRI worked with local telecommunication providers to create simplified disaster alert interfaces for use by visually impaired residents. In another initiative, the institution co-designed offline-accessible apps for multilingual alerts in areas with low network coverage. These projects reflect a commitment not only to technical advancement but to ethical, user-centered innovation that supports the most vulnerable. In addition, DRI collaborates with universities and civil society groups to develop accessible, multilingual, low-bandwidth applications that prioritize user-centered design.

Furthermore, emerging technologies such as drones, augmented reality (AR), and machine learning are being tested for damage assessment, training, and risk communication. In partnership with private sector actors and public agencies, DRI pilots these tools in real-world scenarios and contributes to developing national guidelines on their ethical use. A recent DRI-led project, for instance, involved using drones to assess landslide-prone areas in Hyogo Prefecture and training local students in their operation. This approach not only advances technical capacity but also fosters youth engagement in resilience building. In another case, AR simulations were used in evacuation centers to visualize fire hazards and safe zones, increasing awareness among children and the elderly.

Ultimately, the integration of digital innovation into disaster governance is not simply a matter of efficiency—it is about fostering inclusive, adaptive, and socially responsible systems. DRI's work illustrates the importance of co-design, transparency, and reflexivity in technology deployment. By connecting state-of-the-art tools with grounded community knowledge, Japan's digital risk governance model offers valuable lessons for countries seeking to modernize their disaster strategies while ensuring no one is left behind.

## 7 Climate Change, Social Vulnerability, and Adaptive Resilience

The increasing frequency and intensity of climate-related disasters has prompted a fundamental rethinking of risk governance in Japan. The impacts of global warming—manifested in more violent typhoons, prolonged heatwaves, record-breaking rainfall, and coastal flooding—are not only ecological but deeply social. These events intersect with existing vulnerabilities related to age, disability, migration, and economic inequality, placing disproportionate burdens on already marginalized communities. Consequently, climate adaptation is no longer treated as separate from disaster risk reduction; rather, it is now viewed as an integrated strategy for long-term resilience.

Urban centers such as Tokyo, Osaka, and Fukuoka have begun to incorporate climate adaptation into their urban planning frameworks. These strategies include the development of green infrastructure, such as rain gardens and urban forests, to mitigate heat islands and absorb stormwater; the retrofitting of aging buildings to withstand extreme weather; and the expansion of public cooling centers. Yokohama, for instance, has partnered with academic institutions and civil society organizations to map urban heat risks and design neighborhood-specific interventions. DRI has supported such initiatives by facilitating cross-sector dialogues, compiling international case studies, and promoting participatory planning that centers the experiences of vulnerable populations. It has also led community engagement efforts that combine digital risk visualization with storytelling, enabling residents to understand climate hazards in the context of their lived environments.

In rural Japan, the challenges of climate change are compounded by demographic decline. Mountainous areas in regions like Kyushu, Shikoku, and Tohoku face increasing landslide and flash flood risks, while coastal communities struggle with saltwater intrusion, loss of fisheries, and erosion. These trends have profound implications for food security, livelihoods, and the cultural continuity of rural life. DRI's field programs in these regions emphasize the revival of traditional ecological knowledge, such as *satoyama* (mountain-village ecosystems), and support local efforts to develop adaptive economic models—including eco-tourism, renewable energy cooperatives, and community forestry. Pilot programs have demonstrated how community-managed microgrids, for example, can enhance both energy resilience and environmental stewardship.

One of the most innovative approaches to climate resilience has been the use of scenario-based planning. DRI, in collaboration with local governments and universities, has developed workshops in which community members imagine multiple future disaster scenarios under different climate trajectories. These workshops do not only produce contingency plans—they also help participants articulate values, clarify trade-offs, and imagine more desirable futures. The process often results in the co-creation of locally grounded adaptation strategies, such as relocating essential services, modifying agricultural practices, or redesigning public spaces. In several communities, these workshops have led to the drafting of "resilience charters" that reflect a collective vision for climate adaptation, supported by visual maps and digital storyboards.

Climate resilience also involves addressing systemic inequities. In Japan, elderly women living alone in rural areas are among the most vulnerable to climate-related hazards, yet they are often excluded from formal planning processes. DRI's inclusive resilience model aims to rectify this by partnering with women's associations, disability advocates, migrant support groups, and youth organizations to identify overlooked needs and create space for co-governance. This approach recognizes that resilience is not only about physical infrastructure or climate modeling—it is about power, voice, and justice. By incorporating community facilitators and trained local interpreters into planning sessions, DRI helps ensure that a broader range of perspectives shape adaptation decisions.

Importantly, climate adaptation is now recognized as a global responsibility. DRI plays a role in international knowledge-sharing efforts related to climate-induced displacement, coastal retreat, and climate justice. The institution contributes to policy dialogues within the UNFCCC framework and supports transnational networks of local governments engaged in climate resilience work. Through these efforts, Japan not only responds to its own changing climate landscape but also contributes to a broader vision of equitable and inclusive adaptation. Collaborative projects with Southeast Asian and Pacific Island nations, for example, have involved joint scenario planning, youth-led risk assessments, and shared platforms for climate storytelling. In this way, DRI exemplifies how a nationally grounded institution can serve as a catalyst for global learning and solidarity in the face of planetary risk.

## 8 Conclusion: Reimagining Recovery and Resilience

Japan's evolving approach to disaster risk reduction offers a compelling lens through which to understand the multidimensional nature of resilience. Far from being solely a matter of physical reconstruction or technical efficiency, resilience in the Japanese context is woven through everyday life, social systems, cultural memory, and institutional learning. The experiences of cities like Kobe, Sendai, and Minamisanriku, along with the contributions of institutions like the Disaster Reduction and Human Renovation Institution (DRI), illustrate that recovery is not merely about returning to the status quo ante—it is about reimagining a more just, inclusive, and sustainable future.

The Japanese model highlights the power of community agency. It affirms that local knowledge, mutual aid, and relational networks are not secondary to state-led recovery; rather, they are foundational. This model insists that healing requires public rituals, shared narratives, and the respectful preservation of diverse memories. It also demonstrates that effective risk governance must engage multiple generations, integrate the arts and sciences, and remain sensitive to the lived realities of vulnerable populations.

In confronting the challenges of the twenty-first century—from climate crisis to technological disruption, from demographic shifts to widening social inequality—Japan has embraced a holistic and adaptive framework. DRI's unique integration of emotional recovery, digital innovation, climate adaptation, and community empowerment offers a roadmap for resilience that is deeply human in scope. Its model challenges the prevailing technocratic paradigms by showing that resilience must be imagined, enacted, and sustained through collaboration, empathy, and imagination.

Japan's approach has also shown the importance of continuous learning and flexible governance structures. Initiatives like the so-called "Disaster Tech Dialogue" and scenario-based resilience planning illustrate how Japan is fostering inclusive participation, ethical reflection, and cross-sectoral collaboration. These strategies move beyond reactive risk management to proactive, community-rooted innovation. Institutions like DRI serve not just as repositories of knowledge, but as dynamic platforms for experimentation, critique, and connection—linking hyperlocal experiences with transnational dialogue.

This chapter ultimately argues that disaster recovery is not a singular event but an ongoing process of renewal. It is a social journey grounded in memory, shaped by political will, and animated by collective hope. Japan's case, while context-specific, offers insights that resonate globally. It reminds us that resilience is not about bouncing back—it is about building forward together, with care, creativity, and commitment to justice. The Japanese experience affirms that the future of disaster recovery lies not in isolation, but in shared humanity, inclusive imagination, and an unwavering dedication to transform tragedy into opportunity for all.

## Bibliography

Aldrich, D.P. (2011). The power of people: Social capital's role in recovery from the 1995 Kobe earthquake. *Natural Hazards*, 56(3), 595–611. https://doi.org/10.1007/s11069-010-9577-7.

Aldrich, D.P. (2012). *Building resilience: Social capital in post-disaster recovery*. University of Chicago Press.

Azuma Kagami (吾妻鏡). (1997). Tokyo: Iwanami Shoten.

Bird, P. (2003). An updated digital model of plate boundaries. *Geochemistry, Geophysics, Geosystems*, 4(3), 1027. https://doi.org/10.1029/2001GC000252.

Cabinet Office (Japan). (2012). *White paper on disaster management* (English version).

Cabinet Office (Japan). (2015). *White paper on disaster management* (防災白書).

Cabinet Office (Japan). (2020). *White paper on disaster management in Japan 2020*. https://www.bousai.go.jp/en/documentation/white_paper/pdf/R2_hakusho_english.pdf.

Cabinet Office (Japan). (2022). *White paper on disaster management 2022* (令和4年版防災白書).

Clancey, G. (2006). *Earthquake nation: The cultural politics of Japanese seismicity, 1868–1930*. University of California Press.

Disaster Reduction and Human Renovation Institution (DRI). (n.d.-a). *Institution overview*. https://www.dri.ne.jp/en/overview/.

Disaster Reduction and Human Renovation Institution (DRI). (n.d.-b). *Exhibitions.* https://www.dri.ne.jp/en/exhibition/.

Disaster Reduction and Human Renovation Institution (DRI). (n.d.-c). *International Recovery Forum.* https://www.dri.ne.jp/en/forum/.

Hyogo International Association. (n.d.). *Multilingual mapping boards used at evacuation centers.* https://www.hyogo-ip.or.jp/torikumi/kokusaioryu/documents/comehia_101_eng.pdf.

Hyogo Prefecture Reconstruction Headquarters. (1996). *The record and lessons of the Great Hanshin-Awaji Earthquake* (阪神・淡路大震災の記録と教訓). Kobe: Hyogo Prefecture.

Ministry of Internal Affairs and Communications (Japan). (2021). *Guide to community disaster management plans* (地区防災計画ガイドライン).

Nelson, J. (2003). *Enduring identities: The guise of Shinto in contemporary Japan.* University of Hawaii Press.

Nihon Shoki (日本書紀). (1994). In *Shinpen Nihon Koten Bungaku Zenshū.* Tokyo: Shogakukan.

Reader, I. (1991). *Religion in contemporary Japan.* University of Hawaii Press.

Shaw, R. (2014). *Community practices for disaster risk reduction in Japan.* Springer.

Shaw, R., & Goda, K. (2004). From disaster to sustainable civil society: The Kobe experience. *Disasters*, 28(1), 16–40. https://doi.org/10.1111/j.0361-3666.2004.00241.x.

United Nations Office for Disaster Risk Reduction (UNDRR). (2015). *Sendai framework for disaster risk reduction 2015–2030.*

United Nations Office for Disaster Risk Reduction (UNDRR). (2025). *International Recovery Forum 2025.* https://www.undrr.org/event/international-recovery-forum-2025-achieving-resilient-recovery-changing-world.

CHAPTER 8

# Culturally Sensitive Reconstruction after Disasters: a Long-Term Ethnographic Research in Southwest China

*Qiaoyun Zhang*

Drawing on more than a decade of ethnographic fieldwork materials in the ethnic Qiang regions of southwest China, the chapter reflects on the culturally sensitive reconstruction of the affected ethnic minority areas following the 2008 Sichuan earthquake. The 2008 Sichuan earthquake, also known as the Wenchuan earthquake, was a devastating natural disaster that struck Sichuan Province on May 12, 2008. The earthquake had a magnitude of 7.8 Ms and resulted in over 69,000 deaths, 374,000 injuries, and left about 4.8 million people homeless. The Qiang people, historically concentrated in the region, were particularly hard hit by the disaster. Over 30,000 Qiang people died in the earthquake, accounting for 10% of the group's total population. The earthquake destroyed many Qiang villages, including their traditional slate houses and cultural sites.

Despite the devastating loss, the Chinese government launched a comprehensive reconstruction program, with an estimated total cost of around RMB 1,000 billion. Riding on great state aid and national attention, the Qiang welcomed swift and massive reconstruction, with a special focus on infrastructure rebuilding and cultural preservation. On May 22, 2008, then-Chinese Premier Jiabao Wen emphasized in the earthquake-ruined area that Qiang culture must be well protected after the disaster. In the *General Plan for Post-earthquake Recovery and Reconstruction of Wenchuan published by the Chinese State*, it listed "inheritance and promotion of culture and protection of the ecological environment" as a guiding principle. The rescue and protection of Qiang culture became the first official experiment in post-disaster recovery of national culture.

This chapter presents three distinct yet interactive community rebuilding case studies in the affected areas, aiming to reveal the complex dynamics of post-disaster reconstruction and the renegotiation of culture. The relief and reconstruction projects have fundamentally reshaped the Qiang's community characteristics, livelihoods, and cultural identity. The ethnographic fieldwork has been following the transformation of the Qiang villages since 2012. It has witnessed the construction of the first batch of Qiang heritage and disaster tourism destinations, the transformation of village life driven by new economic initiatives, the Qiang intangible cultural heritage (ICH) safeguarding campaign, the fluctuations in the tourism business, and the daily struggles of the Qiang people.

Following the Qiang region's reconstruction over the past 17 years, the chapter argues that "cultural reconstruction" needs to be analyzed not only from the perspectives of different stakeholders but also in consideration of the various stages of reconstruction and recovery. In this chapter, I propose that the Qiang settlements undergo three stages and types of "cultural reconstruction," ranging from the initial "physical cultural reconstruction" to the ongoing "everyday life reconstruction." As an exemplar of "miraculous reconstruction," the new homes of the Qiang people, which incorporated elements of Qiang cultural symbols and Qiang cultural practices, became a model of cultural reconstruction in the eyes of administrative units, planning experts, and many tourists. It is a rapid, measurable, and accountable, yet superficial, creation of a cultural landscape. I call it "physical cultural reconstruction."

After the rapid "physical cultural reconstruction," the Qiang villagers experienced a long but largely unsuccessful period of home reconstruction. What it points to is how to infuse the newly created physical space with the cultural fabric that constitutes the place and the homeland, including human relations, ethical practices, and beliefs about people and land transportation. I refer to this kind of cultural reconstruction as "relationship reconstruction."

Nearly 20 years after the earthquake, our research in Yingxiu, the epicenter of the quake, has revealed that residents continued to live their lives midst an ongoing struggle with the memory of the disaster and its aftermath. Most residents of Yingxiu returned to the earthquake site, and many of them were involved in disaster tourism activities to make a living. My study of the residents of Yingxiu has provided me with a glimpse of the possible final form of "cultural reconstruction," which I term the "reconstruction of everyday life." Such a study considers how the affected people have the courage and resources to go on living with the trauma of death, the memory of bereavement, and the consumption of disaster and culture.

Through long-term fieldwork in the Qiang region and an examination of different stages and forms of "cultural reconstruction" led by various subjects, the research highlights the diverse and often contradictory connotations of "cultural reconstruction." More importantly, the paper suggests that "cultural reconstruction" should consider the endogenous and compassionate rebirth of the affected subject's intertwined social relations, in addition to the rapid and radical transformation of material conditions. Therefore, the focus of planning and implementing "cultural reconstruction" projects should also shift from providing one-time assistance to long-term reconstruction of everyday life.

## 1 Anthropological Investigation on Cultural Recovery

### 1.1 *Cultural Elements and Disaster Response in Anthropological Perspectives*

In the classic work entitled *Catastrophe and Culture: The Anthropology of Disaster*, Oliver-Smith and Hoffman, the core figures of American disaster anthropology, point out the dual natural-social attributes of disasters and suggest that the "vulnerability" of specific groups of people leads to and enhances the social and material destructiveness of physical phenomena and technological failures. More importantly, the "vulnerability" of populations is conditioned by "the geographic location, infrastructure, socio-political organization, production and distribution systems, and ideology of a given society" (Oliver-Smith and Hoffman 2021:1). The goal of disaster governance is to reduce the "vulnerability" of the affected population and increase the "resilience" of their disaster response.

Oliver-Smith (1986, 1999, 2002) discusses the construction and influence of ecology, social organization, political and economic systems, culture, and practices on disasters by analyzing the relationship between "nature, power, and culture." Susanna Hoffman (1999, 2002, 2015) is one of the first anthropologists to study the cultural component of disaster response. In anthropological terms, culture is a "complex whole" (Tylor 1871) and a framework for explaining what it means to be human. Culture encompasses how people structure their societies, including kinship relationships, livelihoods, political systems, and the temporal and spatial planning of food, clothing, shelter, and transportation. It also encompasses the meanings people attribute to their bodies and environments, the mythic interpretations of the world and its origins, and the distinctive technologies and material cultures they develop.

Culture determines the specific interpretations and coping logics of populations about "environment," "power," "vulnerability," and also "disaster." In China, Folklorists Xiaokui Wang (2022) and Xi Zhang (2022) draw on the experience of Japanese disaster anthropology research and propose the concept of "disaster culture." According to Xi Zhang (2022: 91), Japanese scholar Haruo Hayashi defines "disaster culture" as "a cultural disaster prevention strategy that can be seen in the community where a disaster occurs, consisting of values, norms, beliefs, knowledge, technology, and folk traditions shared among residents. It includes what residents should do in terms of disaster prevention, detection of disaster precursors, and response after a disaster occurs." The disaster experience and disaster memory texts of local societies are the pillars for the formation of "disaster culture." Chinese disaster anthropologist Yongxiang Li (2022: 126) recently argues that disaster culture points to "the relevance and symbolic significance of disasters in socio-cultural contexts" where cultural approaches need to be adopted in disaster governance.

Anthropology focuses on discussing the working mechanisms, interactions, and dynamic changes of cultural elements. Specifically, anthropological investigations of cultural mechanisms during disasters primarily encompass the following three components. First, drawing on the "risk culture" theory, researchers have explored the "social generation" of risk perception (Douglas & Wildavsky 1982; Huang and Xiong 2019, 2021), proposing that cultural categorization systems and social institutions determine whether and how populations can recognize present and potential risks. Schneider's (1957) study of Yap Islanders' response to typhoons notes that potential "dangers" in the physical world, such as typhoons, are perceived as "risks" only when the local political economy collapses and belief systems fail such as poor harvests, tribal tensions, and a chief's failures of virtue. The occurrence of typhoon "disasters" is the result of the inability of the political, economic, religious, and technological systems of society to adapt to risks. Studies on the "environmental refugees" in southern United States (Maldonado 2016), the analysis of the stigmatization of the COVID-19 pandemic (Singer et al. 2021), and the investigation of the evolution from "fire hazard" to "fire disaster" in ancient villages in Southwest China (Tang 2015) collectively reveal the social constraints and cultural construction in the process from risk perception to disaster.

Another batch of studies has problematized the misjudgment and misuse of culture in disaster governance measures. Scholars have noted that key cultural elements in disaster situations include individuals' arrangements for living space, their calculations of time, and their use of class, gender, and age as identity resources, as well as community organization and unity, and religious belief norms, among others (Hoffman 2015). However, such cultural practices are often regarded as "backward" notions or beliefs of the affected population. In places like Peru (Oliver-Smith 1986), Greece (Hoffman 2002), New Orleans in the United States (Barrios 2017), Haiti (Schuller 2012), and India (Fortun 2001), failed disaster governance often stems from the deliberate neglect and erroneous transformation of local knowledge and social relationship networks such as kinship systems, neighborhood relations, occupational characteristics, and living spaces by expert knowledge or technical rationality. The theory of "disaster capitalism" critiques the destructive impact of economic rationality, based on cost-benefit principles, on disaster response and decolonization processes in the Global South (Schuller 2008, 2012, 2016). The culture of some ethnic groups is considered "frozen tradition" (Fabian 1983). Post-disaster rescue and protection of local culture are often seen as obstacles to modern development (Barrios 2013, 2015). Moreover, culture, especially some intangible cultural heritage items, is commodified into handicrafts and stage performance projects in post-disaster reconstruction.

Last but not least, anthropological studies on the relationship between humans and nature have explored the emergence of diverse cultural practices, meanings, and values, as well as their ecological relevance (Biersack 1999; Ingold 2000). The concept of culture helps state institutions define the cultural elements of affected populations in post-disaster situations (Bennett 2015; Hastrup 2011). Ethnographic investigations of the 3.11 earthquake in Japan (Gill 2013), the 2004 tsunami disaster in Sri Lanka (Gamburd 2013), and the 2008 Wenchuan earthquake in China (Zhang and Tang 2018) have respectively discussed the cultural connotations of "home," "landscape," "place attachment," and "sense of place." They propose that post-disaster reconstruction should be a dual process, rebuilding both the physical world and cultural relationships, as well as local communities and the broader society.

### 1.2 *Disaster Anthropology towards Chinese Practice*

Chinese disaster anthropology draws on and transcends the Western political economy theoretical framework, as well as its academic goals of critiquing capitalism and colonialism, to independently interpret and evaluate the unique institutional and cultural patterns of China's disaster response. Chinese disaster anthropology has, from the outset, emphasized the significant importance of cultural protection for economic development, cultural identification, and harmonious co-prosperity in ethnic regions. The research areas are primarily located in disaster-prone regions of southwest China, including Yunnan, Guizhou, and Sichuan. The types of disasters they have dealt with are principally natural disasters, including floods, earthquakes, landslides, and droughts. The Chinese experience, led by the state and aimed at promoting modern development and cultural-ecological protection, offers a unique and valuable inspiration for the world's research on the cultural mechanisms of disaster response.

In terms of theoretical innovation, inspired by the "ontological turn" in anthropology and combined with Chinese political anthropologist Xiaoyang Zhu's (2015) interpretation of "terrain," (dishi; 地势), Chinese disaster anthropologists Zhang Yuan and Tang Yun (2020, 2022) adopt the concept of "dwelling habitat" (*renwen shengjing*; 人文生境)—the growth activities and interactive realms of all beings in a specific spatio-temporal domain—to explain disaster and risk scenarios. The concept of "dwelling habitat" reminds researchers not only to consider sociocultural mechanisms related to humans but also to pay attention to the agency of various non-human factors, such as animals, plants, wind, water, and soil. Social resilience should be shaped around the concept of the "dwelling habitat." Similarly, anthropologists A.J. Faas and Roberto Barrios (2020) also warn us, in their discussion on the anthropological research agenda for COVID-19, to attend to multispecies interactions.

On methodological exploration, Chinese disaster anthropologists have argued that research on death and destruction must "orient to the living world" (Zhang and Tang, 2011) and that disaster ethnography needs to examine local life experiences and knowledge landscapes (Zhang and Ma 2013). Xiaokui Wang (2022: 83–84) proposes that "disaster folklore ethnography" (*zaihai minsuzhi*; 灾害民俗志) should be a key methodology for disaster research. The "disaster folklore ethnography" draws on the theory of "disaster culture" and focuses on investigating the common-sense system underlying disaster events. The "common-sense system" refers to local knowledge about production, life, and living that has been passed down through generations, is "collectively shared, and is relatively stable, tied to specific regions" (Wang *ibid*). After generations of practice, the knowledge has become an internalized "common-sense system," akin to physiological "instinct," and can provide people a sense of security and comfort. The "common-sense system" contains a significant amount of tacit knowledge based on physical sensations, which cannot be described using systematic concepts (Wang *ibid*).

The key task of disaster ethnography is to examine the tacit knowledge embedded in the "common-sense system," as it critically complements scientific knowledge widely upheld in disaster recovery research and practice. As shown in Wang's research on the 2011 Tōhoku earthquake and tsunami in Japan (*ibid*), residents of the Kesennuma area in Eastern Japan, the most severely affected area, rejected a scientifically based tsunami defense plan due to their deep emotional and cultural connections to the sea. Instead, they chose to coexist with the risk, valuing their way of life and cultural heritage over absolute safety. The case study illustrates the tension between scientific rationality and local "common-sense systems" in post-disaster reconstruction, offering insights into the balance between life and risk in disaster-prone areas.

Closely investigating disaster recovery in the Qiang region of Southwest China following the Sichuan Earthquake, this research extends beyond the traditional political economy framework of Western disaster anthropology to examine the cultural mechanisms and disaster coping systems in the Chinese context. At the same time, it explores the applicability of anthropology's particular conceptualizations of the "dwelling habitat" of disaster scenarios and the "common-sense system" behind disaster events through specific ethnographic studies, explaining the basic elements of the cultural mechanisms for coping with natural disasters.

## 2 Ethnographic Site and Methodology

Longxi Township is located in Wenchuan County, within the Aba Tibetan and Qiang Autonomous Prefecture of Sichuan Province. The township is situated in a remote mountain valley, characterized by its rugged terrain and limited arable land. Before the earthquake, Longxi was economically underdeveloped, with the majority of its population relying on subsistence agriculture and small-scale animal husbandry. The Qiang people, who comprise over 95% of the township's population, had long lived a relatively isolated and challenging life before the earthquake. When villagers first introduced me to their pre-earthquake lives in 2012, I was deeply moved by the fact that their lives had hardly improved compared to the records of David C. Graham (1958), the first Qiang scholar from the United States to conduct field research in Longxi between the 1920s and 1940s.

Longxi's reconstruction has drawn the attention of many scholars and reporters, mainly for the following three reasons (Yang and Jiang 2012; Wu 2012). First, Longxi was heavily damaged by the earthquake, with nearly all of its buildings, infrastructure, and public services destroyed. Second, the township was selected as an exemplary post-disaster recovery and reconstruction site in Wenchuan, where a large-scale reconstruction program aimed at rebuilding the township, preserving Qiang cultural heritage, and promoting tourism proceeded swiftly and rigorously. This reconstruction effort was not merely about physical rebuilding, but also about cultural revitalization, which led to the development of cultural tourism as a new economic strategy for the region. Last but not least, the township was also home to several generations of Qiang *shibi*, the ritual specialists, who were custodians of local myths, history, and spiritual practices. Traditional rituals and beliefs have continued to guide the villagers' lives. Longxi was renamed as "The Qiang People's Valley" after the reconstruction.

The overall aim of the longitudinal study is to investigate the complex dynamics of the culturally sensitive recovery of an ethnic minority region. In the process, various forms of tourism have played a central role, while the conflicts and negotiations between tourism development and community rebuilding are increasingly intense. Longxi's long-term fieldwork witnesses the rise and fall of dark tourism and heritage tourism promoted by the overwhelming state. My recent fieldwork in the nearby Yingxiu township, the epicenter of the earthquake, explored the heart-wrenching question of how different stakeholders perceived and experienced death and haunting while visiting and living in the earthquake-ruined areas and situ-reconstructed homes. The ethnographic fieldwork in Longxi lasted for more than a decade. For this article, I conducted 50 structured interviews with Longxi villagers, government officials, planners, experts, and scholars to gather their perceptions and practices regarding earthquake relief and reconstruction. More than 100 informal interviews were conducted throughout the fieldwork, focusing on the transformation of villagers' lives. Two surveys were conducted among 40 Dongmen villagers to assess their perceptions of local disasters and the reconstruction efforts. All ritual and festival practices were visually documented when permitted. The field research in Yingxiu lasted for two months in 2023, during which participant observation and structured interviews were conducted to gather different interviewees' ideas about Yingxiu's rebirth and its role in dark tourism.

## 3 Reconstruction of Physical Space for Tourism Development

As early as 2012, when I first arrived in Longxi, I noticed two very conspicuous banners hanging from the solar-powered street lamp posts on both sides of the newly repaired asphalt roads. The left-side ones read, "Chinese Qiang People's Valley; Mysterious Longxi Gully" (*zhongguo qiang ren gu, shenmi longxi gou* 中国羌人谷，神秘龙溪沟). The ones on the right side read, "Witness the Miraculous Reconstruction; Appreciate the Rise of the Great Love" (*jianzhen chongjian qiji, ganshou da'ai jueqi* 见证重建奇迹，感受大爱崛起). As the banners indicated, the post-disaster reconstruction in the Qiang villages was officially promoted as a "miraculous" case, an infrequent occurrence in global disaster response and management projects.

The "miraculous reconstruction" is first and foremost characterized by its swiftness and massive scale. According to the related governmental reports, under the strong support of post-disaster restoration and reconstruction, the economic development of the 39 hard-hit and extremely hard-hit counties and cities in Sichuan was accelerated, and the basic living conditions and the level of economic and social development in those areas reached or exceeded the pre-earthquake level in 2009. In Longxi alone, all affected villagers were able to return to their newly constructed or repaired homes by the time of the incoming Chinese New Year, just 8 months after the earthquake. Public facilities, including a new township hospital, primary school, governmental buildings, and bus stations, were rebuilt with the technical and financial support coordinated by the Chinese state. Infrastructure, including cross-township asphalt roads, tap-water pipelines, garbage stations, and solar street lamps, was first introduced to this remote township. Several of Longxi's villages were also planned to be rebuilt and developed as Qiang cultural tourist sites.

There is another layer added to the swift physical reconstruction of the Qiang villages to make the rebuilding "miraculous:" the cultural touch. In Longxi's reconstruction, there was a heavy emphasis on cultural recovery. However, such reconstruction was initially limited to the physical reproduction and display of Qiang culture. I have analyzed the process as "political transplanting" (Zhang 2012), where various Qiang cultural elements, both local and imagined, were transplanted into the Longxi villages to rebuild them as "Qiang people's Valley." For example, the transplanted scenic spots filled Longxi's Dongmen village, which was transformed into a tourist destination by the reconstruction plan. In the new welcome plaza, a one-meter-high brick sacrificial tower was erected. At the center of the village, a 5-storey Qiang watchtower was built. Neither of the architectures was original in Dongmen. They were "transplanted" to Dongmen because one symbolized the Qiang's animistic sacrificial practice and the other condensed the Qiang's history of defense and migration to Sichuan. Meanwhile, the new reinforced concrete houses were covered with cement sheets that look like stone chips to "restore" the Qiang people's tradition of building homes with stones not to mention the fact that most of the village's buildings were painted with brown totem-like patterns based on sheep's heads on their facades to remind visitors of the nomadic lifestyle of the Qiang people.

Arguably, the "physical cultural reconstruction" projects can be seen as a form of cultural spectacle that achieves multi-layered goals by reviving and preserving the unique artistic traditions of the Qiang people. On the one hand, such projects were designed to showcase the Qiang culture in a way that aligned with the nationalist agenda and development goals. The swift and massive reconstruction demonstrates the Chinese state's overwhelming power and determination in "fighting against the earthquake and relieving people from the disaster" (*kangzhen jiuzai* 抗震救灾) (Edgerton-Tarpley 2014). The warlike terminology used in disaster relief reflects the significant role of powerful disaster relief in demonstrating the state's authority and the nation's unity and stability, especially in China's ethnic regions.

On the other hand, the cultural preservation planning also paved the way for tourism development for the reconstructed villages. The reconstruction planning echoed a popular development scheme that began with China's Opening-up Policy in the late 1970s. It was vividly called "using culture to set up the stage for the economy to sing" (*wenhua datai, jingji changxi* 文化搭台，经济唱戏), where culture was used as an engine and commodity for economic and other development. Researchers working on Chinese heritage tourism development since the early 1990s have also observed the changing state policy which marked the "cultural turn" in the ethnic areas' development strategies, in which "cultural resources (including repackaged and reinvented cultural heritage) have come to be viewed as possessing equal if not greater economic potential than traditional factor endowments typically marshaled by local governments in their development planning" (Oakes 2006:14).

The physical cultural reconstruction plan in Longxi was designed to appeal to urban, middle-class, predominantly Han tourists' nostalgic imaginings of rural, ethnic life (Zhang and Barrios 2017). This led to the exoticization of Qiang villagers and the commodification of their culture. State planners and tourism agency project managers reduced a complex culture to a simplified commodity, transforming it into a tourism product. Examples of cultural grafting include the construction of two Qiang-style "stone" watchtowers, made of concrete but decorated with imitation stone pieces. The *shibi* practice, which was not originally part of Dongmen's culture, was prominently represented through performances and sculptures. The transformation of the village into a tourist destination resulted in significant changes to the local economy and social structure. Much of the village's limited farmland was expropriated to build tourism-related facilities, leaving villagers vulnerable to the financial vagaries of the tourism industry. The villagers were not consulted or included in the planning of reconstruction projects, and their compliance was expected and sometimes forced. The villagers found themselves in a situation where their culture was being exploited for economic gain, resulting in a profound sense of cultural alienation.

## 4 Relationship Rebuilding for Reconstructing Home

As discussed in the previous section, the Chinese state's approach to post-disaster reconstruction in Longxi prioritized efficiency and modernization over sustainable and sensitive community recovery. The key reason leading to the unintended consequence of the failed home making lies in the fact that while housing reconstruction focuses on the physical rebuilding of structures, home reconstruction involves the re-establishment of social, cultural, and emotional connections to a place. Social scientific research has long carefully distinguished between space and place, where place-making is analyzed as the creation of various geophysical and socio-economic relations that sustain both material and spiritual needs and hopes of the people (Douglas 1991; Gupta and Ferguson 1992; Low and Lawrence-Zuniga 2003). Similarly, in post-disaster reconstruction, ethnographic research has highlighted the distinct differences between housing construction and homemaking. In many reconstruction projects, redevelopment is often equated with the reconstruction of housing. It employs a top-down approach to solving livelihood problems from the perspectives of planning and efficiency, with its primary purpose usually being to demonstrate planners' ability to mobilize resources and manage the masses for a type of "high modernist" development (Scott 1998: 1–8).

Anthropological research, on the other hand, is more concerned with reconstructing local spiritual and cultural life. It explores how individuals affected by environmental challenges can reconfigure various ecological resources and interpersonal networks to create and organize their ideal lives, thereby building a place that can be called home (Barrios 2011 and 2014; Brun and Lund 2008; Hoffman 2016). Bourdieu's description of habitus can effectively explain the definition of home at both spiritual and cultural levels. Hoffman writes (2016: 47), "At the heart of a community lies habitat, but habitat is not merely housing. Beyond shelter, habitat involves the paths of human movement, sites of association, comfort of familiarity, roots of identification, sacred space both individual and collective, sight, sound, and smell."

In Longxi's historic village, E'er Village, my colleagues and I conducted long-term research to investigate the multi-layered dynamics of home-making among the villagers since the earthquake (Zhang et al. 2024). The research was inspired by a villager's comment during an interview, in which he lamented, "Home is not like home, the village is not like a village, and people are not like people" after the reconstruction. Somewhat ironically, this interview took place in front of rows of newly renovated, government-funded, reinforced concrete houses in E'er. The villagers referred to these houses as "concrete boxes" (*shuini hezi* 水泥盒子) rather than homes. Examples of housing reconstruction, known as the "concrete boxes," were merely physical structures that provided shelter and served as symbols of the rapid restoration of basic living conditions. Villagers explained that building concrete houses in the designated areas was primarily due to governmental funding policies, which required them to rebuild such qualified and subsidized houses promptly. Often driven by state-led initiatives, housing reconstruction projects aim to quickly restore physical spaces, ensuring that affected populations have a place to live. These projects are typically characterized by standardized construction methods and materials, prioritizing speed and scale over cultural sensitivity and context.

Housing reconstruction is often tied to broader economic and political objectives, including urbanization, economic development, and national unity. The state's role in these projects is to ensure that the reconstruction aligns with national development plans and policies. According to the reconstruction plan for Wenchuan County, E'er Village was to be transformed into the "origin of Qiang shibi culture" after the earthquake, showcasing the religious tradition and authentic living conditions of the Qiang people, in support of Longxi's new branding—"Qiang People's Valley." *Shibi* were the ritual specialists and masters of the local Qiang myths and history, as well as monitors of village rituals and consultants for villagers' major life decisions. E'er village has hosted several multi-generational *shibi* families. The masters not only monitored the most essential Qiang ritual festival—the Qiang New Year—for generations, but also practiced the Qiang sheep-skin dance, mortuary ceremonies, and traditional medical practices. The Qiang New Year and Qiang Sheep-skin Dance are both listed in the Representative List of Chinese Intangible Cultural Heritage. The former was added to the UNESCO Representative List of Intangible Cultural Heritage in 2024.

From the very beginning, E'er village was built with cultural presentation and tourism development in mind. The houses in the town were built along the main roads of the village, using the same cement and stone sheets as those in Dongmen Village, to showcase the Qiang's "ancient tradition" and facilitate tourists' visits and stays in the village. The sacrificial pagoda and watchtower in the town were also rebuilt, but were transplanted to the newly constructed public square, a convenient destination for tourists to gather and visit. During holidays and tourist promotional periods, the *shibi* and their apprentices were invited by the government to demonstrate Qiang customs in E'er or Dongmen, featuring performances of wine toasting, rituals, and the sheep-skin drum dance.

Our ethnographic research revealed a profound epistemological and political-economic divide between development-oriented housing reconstruction and the cultural and social aspects of homemaking. Home reconstruction extends beyond the physical rebuilding of houses to encompass the reestablishment of social networks, cultural practices, and emotional connections to a place. It involves restoring community solidarity, social cohesion, and the sense of belonging that defines a home. In particular, the villager who commented on "home not like home" elaborated that the dilution of *renqing* (human feeling 人情) was the main reason for it. *Renqing* refers to the human feelings and social relationships that are nurtured through various practices of house building and interaction among the Qiang people. It is a culturally specific concept that reflects the reciprocal gift-giving relationships among close kin, friends, and acquaintances, which build one's social network, status, and authority in a familiar setting. *Renqing* is maintained through practices such as *zuo renqing* (doing someone a favor) and *qian renqing* (owing someone a favor), which involve both emotional attachments and ethical (and sometimes economic) responsibilities to sustain interpersonal relations.

To analyze what composed of the villager's conception of *renqing* in E'er, we studied the mutual aid labor form of labor exchange in the economic life of villagers in Longxi Township, examined the long-standing in-law relationship between villages, and derived the socio-economic reasons why the Qiang people preferred to build their houses in the form of clusters. We also found that the construction of homes by the local Qiang people was a long-term, mutual-help, and ritualistic process, and that building a family house was also a process of establishing and confirming kinship, good neighborliness, and human-spiritual relations.

However, in the process of rapid reconstruction and tourism development, the villagers' economic, social, and faith-based relationships, which were previously connected by the construction of their homes, had been disrupted. Villagers had to rebuild their houses scattered across the area in a short period, according to the planned sites, where there was neither mutual assistance in labor nor prayers from the *shibi*, but rather competition and antagonism. Villagers were often at odds over building materials or better sites, resulting in their separation and growing geographical and psychological estrangement. This was perhaps the main reason why the villagers lamented that *renqing* faded.

During our subsequent fieldwork, we heard another saying from an old *shibi*. He noted that Longxi could not return to its former state because the *qi* of Longxi had vanished. *Qi* (气) described the good fortune and sacredness of the land fostered by disciplined, collective spiritual worship. It was closely related to the Qiang's geomantic practices and represented the spiritual connection between the people and their land. *Qi* was nurtured through rituals and ceremonies that honor the spirits and ancestors, and it played a crucial role in the Qiang's understanding of their environment and their place within it. The Qiang villagers' desire to remake *renqing* through proper house-building practices was interrupted by the state-led reconstruction projects. Rituals and worship practices, which were central to maintaining *qi*, were abandoned in favor of rapid rebuilding. For example, the careful selection of "good land" for rebuilding ceremonial worship pangoda and watchtower, guided by *shibi* and local knowledge, was neglected in favor of rapid construction and tourism development. This neglect resulted in a loss of *qi*, as the new structures were not properly consecrated and were later destroyed by a mudslide. The mudslide itself was also regarded as an indication of the loss of *qi* and *renqing*. Moreover, rituals and ceremonies that were once central to maintaining *qi* were transformed into tourist performances, stripping them of their spiritual significance. The ritual process is a crucial means of integrating the local community. When severe and intimidating rituals were packaged for tourism, villagers lost the opportunity and access to nourish the *qi* of Longxi through such collective practices.

Compared to our research on the cultural transformation of Dongmen village in the early stages of reconstruction, our ongoing investigation in E'er village reveals a deeper level of cultural reconstruction, specifically the reconstruction of the spiritual and cultural life of the place. Starting from the difference between housing reconstruction and homemaking, and by analyzing the cultural connotations and political dynamics of the concepts of *renqing* and *qi*, the research highlights that for the Qiang, homemaking was not merely about constructing physical shelters, but also about rebuilding social and cultural connections. *Renqing* and *qi* were essential elements in this process. Through *renqing*, the Qiang established and maintained social networks that provided mutual support and assistance. These networks were vital in times of disaster, as they helped communities come together to rebuild and recover. *Qi*, meanwhile, reinforced the spiritual bond between the Qiang and their land, providing a sense of continuity and a more profound sense of belonging. Therefore, the essence of what the article refers to as "relationship reconstruction" lies in the reweaving of various culturally significant relationships, including kinship, neighborhood, ritualistic practices, and supernatural beliefs. Such relationships enable individuals to unite in solidarity and live sustainably in areas previously prone to disasters. *Renqing* and *qi* can be considered forms of agency that are more effective than tourism-driven development in building solidarity, creating more sustainable lives, and envisioning better futures. Only reconstruction that can help communities reorganize these relationships is culturally sensitive, and thus can serve on-the-ground, ongoing, and constructive functions.

## 5 Reconnecting the Past for the Reconstruction of Everyday Life

Over the years of the fieldwork, I have visited numerous reconstructed villages, but to be honest, I have rarely visited the earthquake sites. For one thing, I felt that I had learned the details of the earthquake so well that I did not need to see the actual sites. More importantly, I did not know if I dared to face such a massive trauma of life and material. In 2018, ten years after the earthquake, I seized the opportunity of an academic conference to revisit the ruins of the old Beichuan county seat, which had been preserved intact. The old Beichuan county seat was the center of Beichuan County, the only Qiang autonomous county in China, at the time of the earthquake. The county seat, situated in the seismic zone, was almost destroyed in a matter of minutes due to the quake triggering large-scale landslides and geological subsidence. Not a single house was left intact in the county. The most heartbreaking aspect was that Beichuan Middle School, located in the old county seat of Beichuan, completely fell down in the earthquake and was subsequently buried by the collapsed mountain behind it, leaving only a flagpole and a basketball hoop on the ground.

At the site, I saw a white banner, over 10 meters long, written in black letters, hung over the school's ruins, which were preserved intact. The banner's headline read, "Sorrowful Condolences to All the Teachers and Students of the School," and the two sentences at the beginning of the banner read, "Dear son, are you okay? Mom misses you!" According to a media report (Ding 2018), the mother of a student buried in the ruins insisted on visiting the school site regularly after the earthquake to hang a banner with similar content. This mom said that she put all the things she wanted to say to her son in the banner and left it for him to read. This time, the newest banner overlapped the old ones. The son's photo was printed in color, and the banner ended with the mother's cell phone number. According to the same news report, the mother never changed her cell phone number, fearing that one day her son would return and be unable to reach her.

This tear-jerking scene prompted me to consider another crucial issue related to long-term, culturally sensitive post-disaster recovery: how those who remained in the affected areas and survived the earthquake can rebuild their daily lives midst the experiences and memories of the disaster and its aftermath. Inspired by the unforgettable experience of visiting the old county seat of Beichuan and my collaboration with a scholar of tourism, I conducted a study on haunting, death, memory, and everyday economics in Yingxiu town, the epicenter of the earthquake, in 2022 (Zhang and Zhong 2024). After the quake, Yingxiu was rebuilt on its original site. In addition to preserving the ruins of Xuankou Middle School, a newly constructed cemetery for the victims and a memorial hall were built at the epicenter of the earthquake. Most of the villagers who survived the disaster relocated to the original site and resided in new houses constructed with government funding. It became the site where everyone faced a lingering death head-on. It was also the site where most of the survivors had to make a living from, with, and for the catastrophic deaths.

Hauntology, as conceptualized by Derrida (2006), refers to the presence of the past in the present, often through the concept of specters. The specters represent the invisible, intangible, or ghostly phenomena that linger in contemporary work, shaping our understanding of places and events. In the context of memorization, hauntology can help us recognize how the past continues to influence the present, even if it is not immediately visible or acknowledged.

The memorization of different kinds of deaths can play a significant role in reconstructing the everyday lives of individuals and communities by reconnecting them to the past. The process involves acknowledging and remembering the deceased, understanding the circumstances of their deaths, and finding ways to integrate these memories into the present and future. It consists of the creation of spaces, practices, and narratives that honor and commemorate the deceased. It can take many forms, from physical monuments and museums to cultural practices and personal rituals. By engaging with these memorials, individuals and communities can reconnect with the past and find meaning in their experiences.

Dark tourism, which involves visiting sites associated with death, tragedy, and suffering, can play a significant role in the memorization process (Lennon and Foley 2000). It provides individuals with the opportunity to engage directly with the past, fostering a deeper understanding and connection to the events and people involved. In Yingxiu, dark tourism has been carefully managed to create a space where visitors can reflect on the past and connect with the present. The preserved ruins, museums, and other memorial sites served as powerful reminders of the earthquake's impact, while also providing a space for learning, education, and commemoration.

In Yingxiu, we observed various forms of memorialization for different types of deaths, reflecting different interpretations of the concepts of "martyrs," "spirits," and "ghosts" in official narratives, among tourists, and local villagers. The spirits of the martyrs, such as the military helicopter crew led by Guanghua Qiu, were memorialized through official sites like the epicenter museum and public cemetery. The state's representation of these deaths reframed them as heroic sacrifices, transforming the tragedy into a narrative of national unity and resilience. It served to neutralize the negative aspects of the disaster and instead highlighted the positive aspects of sacrifice and salvation.

The deaths of ordinary individuals, such as the students and staff of Xuankou Middle School, were memorialized in a more subdued manner. The preserved ruins of the school served as a powerful reminder of the lives lost, while the museum's exhibits and narratives attempted to provide a sense of closure and remembrance. The "Flower of Life" artwork, made from the ashes of two victims, was a poignant example of how the deceased can be remembered in a way that is both respectful and transformative.

The local villagers' engagement with invisible beings and spirits represented a more personal and intimate form of memorization. After the reconstruction, most residents of Yingxiu derived their livelihoods from activities related to earthquake tourism. Residents living in houses on the street converted their homes into hostels, restaurants, or stores. Those who were unable to do so either rented government-designated stalls to sell their goods, became full-time tour guides, or purchased vehicles to transport guests or goods. In any case, their daily lives were closely tied to the disaster that had cost them relatives and friends. Their everyday conversations and lives were also filled with haunting presences. They clearly remembered how many corpses were parked in the place where a new park was built, and could point out which family's relative the ghostly figure the child said he saw belonged to. When asked if they were afraid of ghosts coming for them, they could also give a tearful response, "I wish they would." By acknowledging the presence of these spirits in their everyday lives, the villagers were able to maintain a connection with the past and find a sense of continuity and belonging.

Reconnecting with the past is an indispensable part of reconstructing everyday life. In Yingxiu, we found that all people involved created their meaning and understanding of the disaster by engaging with their past experiences and the things they had lost. Honoring the memory of heroic martyrs helps us understand the greatness of sacrifice and the fearlessness that accompanies death. The memory of pure souls, as embodied in the "Flower of Life" artwork, helps us remember the deceased and cherish the people who came before us. The daily engagements and narratives of the invisible lives allow individuals and communities to come to terms with their losses and move forward with their lives, embracing a future towards death (*xiangsiersheng* 向死而生). In this regard, I propose that the meaningful cultural recovery lies in the reconstruction of everyday life. Such a sensitive reconstruction would allow for the constant encounter with death, the lasting storytelling of loss, and the economics of grief. It would also enable survivors and those concerned to intimately commemorate the disaster and continue the future with close ties to the past.

## 6 Conclusion

Based on more than 10 years of research on the post-earthquake recovery and reconstruction in the Qiang region of Wenchuan, China, this chapter attempts to systematically explain and demonstrate the multiple meanings and contested practices of "cultural reconstruction." We observe that "cultural reconstruction" is understood and practiced differently at various stages of recovery and reconstruction, as well as by different stakeholders.

In the government-led post-disaster emergency and recovery phase, which emphasized efficiency and dominance, "cultural reconstruction" was often simplified or even downgraded to a clumsy, biased, and commodified physical recreation of Qiang material culture, focusing on architectural styles and art forms. Such a tourism-oriented physical cultural renovation, in the view of villagers and tourists, felt a little "fake." Analyzed from the perspective of cultural anthropology, the reason why people thought that the culture was "fake" is that it was detached from a culture's context, which includes environmental, geographic, social, and humanistic factors that allow culture to take root in a place, and thus have the ability to explain and guide life.

Therefore, in the research of E'er village, we point out the important connotation of "cultural reconstruction"-the reconstruction of "relationships." "Relationships" in this context refer to the dynamic relationships between relatives, neighbors, and villages that organize and maintain the economic life and social identity of the Qiang people. They also involve the interaction of human and non-human existence, woven into the belief in the eternal continuity of the Qiang people's patronage and identity with their gods, ancestors, and forefathers. These "relationships" turn these villages into homes, a place of unending human feelings (*renqing*) and good fortune (*qi*).

After exploring the concept of "cultural reconstruction," ethnographic fieldwork suggests that anthropology's concern with culture needs to be revitalized. We need to recognize that the ultimate significance of culture lies in establishing and guiding daily life, where its form and role are also shaped. One of our recent investigations forced us to confront the long-lasting effects of disaster and death, exploring how disaster-affected populations can continue to live desirably in the place where the disaster originally occurred, while wrestling with the experience and memory of death. By examining the categorization, representation, and memory of different deaths in various populations, we demonstrate that the prerequisite or basic "script" for the reconstruction and continuation of everyday life is the interpretation and acceptance of such sudden and tragic events as disasters. Chinese people often say that "there will be good fortune after a great disaster." This unique Chinese wisdom reflects the bravery and determination inherent in the Chinese philosophy of life, which is to "live towards death." With the belief that we must die, we accept the challenges and traumas that everything throws at us, such as disasters, bereavement, and loneliness. But also because of the curiosity of life, we can carry all kinds of ghosts and spirits with us as we continue to walk the one-way road of life. Daily life is a process of finding hope in despair and meaning in the mundane. It is this kind of faith, determination, and reverence that we need to rebuild.

Admittedly, ethnographic fieldwork on "cultural reconstruction" needs to continue. One limitation of this study is that it primarily focuses on a single region and lacks a comparative cultural perspective. In addition, the discussion of cultural matters in this study is not comprehensive enough. My colleagues and I plan to continue discussing cultural restoration and change more holistically and dynamically by examining more cultural aspects, including Qiang songs, dances, legends, skills, and festivals and ceremonies. We strive to provide more Oriental wisdom for the "toolbox" of a desirable human life through discussions of life's radical changes in cross-temporal and cross-cultural exchanges and scenarios.

## Bibliography

Barrios, Roberto E. 2011. "'If You Did Not Grow Up Here, You Cannot Appreciate Living Here': Neoliberalism, Space-Time, and Affect in Post-Katrina Recovery Planning." Human Organization 70, no. 2: 118–27.

Barrios, Roberto E. 2014. "'Here, I'm Not at Ease': Anthropological Perspectives on Community Resilience." Disasters 38, no. 2: 329–50.

Brun, Cathrine, and Ragnhild Lund. 2008. "Making a Home during Crisis: Post-tsunami Recovery in a Context of War, Sri Lanka." Singapore Journal of Tropical Geography 29, no. 3: 274–87.

Derrida, Jacques. 2006. *Specters of Marx*. Routledge.

Ding, Wei. 2018. Beichuan Mama Shinian 30 tiao Hengfu Xiegei Tiantang de Erzi: Mama Hao Xiang Ni [Beichuan mother ten years 30 banners written to the son in heaven: mom misses you]. https://news.sina.com.cn/s/2018-05-08/doc-ihacuuvu7359774.shtml, accessed May 8, 2025.

Douglas, Mary. 1991. "The Idea of a Home: A Kind of Space." *Social Research* 58(1): 287–307.

Edgerton-Tarpley, Kathryn. 2014. "From 'Nourish the People' to 'Sacrifice for the Nation:' Changing Responses to Disaster in Late Imperial and Modern China." *The Journal of Asian Studies* 73(2): 447–469.

Graham, D. Crockett. 1958. *The Customs and Religion of the Ch'iang*. Smithsonian Miscellaneous Collections 135(1).

Gupta, Akhil, and James Ferguson. 1992. "'Beyond Culture': Space, Identity, and the Politics of Difference." *Cultural Anthropology* 7(1): 6–23.

Hoffman, Susanna M. 2016. "The Question of Culture Continuity and Change after Disaster: Further Thoughts." *Annals of Anthropological Practice* 40(1): 39–51.

Lennon, J. John and Malcolm Foley. 2000. *Dark tourism*. London, UK: Thomson.

Low, Setha M., and Denise Lawrence-Zuniga. 2003. "Locating Culture." In *The Anthropology of Space and Place: Locating Culture*, edited by Setha M. Low and Denise Lawrence Zuniga, 1–48. Hoboken, NJ: Wiley-Blackwell.

Oakes, Tim. 2006. Cultural Strategies of Development: Implications for Village Governance in China. *The Pacific Review* 19(1): 13–37.

Scott, James C. 1998. *Seeing like a State: How Certain Schemes to Improve the Human Condition Have Failed.* New Haven, CT: Yale University Press.

Wu, Shan. 2012. *E'er Cun: Bei Chonggou de "Mingxing Cun"* [*E'er Village: A Reconstructed "Star Village"*]. Nanfang Dushi Bao [South China Metropolitan Newspaper], May 12. https://news.sina.com.cn/c/2012-05-12/045324405092.shtml, accessed May 6, 2025.

Yang, Zhengwen and Bin Jiang. 2012. *E'er Cun: Yuanjian Zhudaoxiade Zaihou Chongjian Moshi* [*A'er Village: Post-disaster Reconstruction Program Dominated by Aid Construction*]. Wuhan: Huazhong Keji Daxue Chubanshe [Huazhong University of Science and Technology Press].

Zhang, Qiaoyun. 2012. Phoenix out of the Ashes: Convergence of Disaster and Heritage Tourism in Jina Qiang Village, Sichuan Province, China. *Practicing Anthropology* 34(3): 24–28.

Zhang, Qiaoyun, and Roberto E. Barrios. 2017. "Imagining Culture: The Politics of Culturally Sensitive Reconstruction and Resilience-Building in Post-Wenchuan Earthquake China." In *Responses to Disasters and Climate Change: Understanding Vulnerability and Fostering Resilience*, edited by Michèle Companion and Miriam Chaiken, 93–102. Boca Raton: CRC Press.

Zhang, Qiaoyun and Shuru Zhong. 2024. "The politics of spectrality in earthquake ruins: remaking dark tourism in hauntology." *Tourism Geography*. Published online. https://doi.org/10.1080/14616688.2024.2358330.

Zhang, Qiaoyun, Guanli Zhang, Wen Huang, Yun Tang. 2024. "The Social Production and Political Potency of Home: Contentions between Housing Reconstruction and Place Making among the Chinese Qiang after the 2008 Sichuan Earthquake." *Positions: Asia Critique* 32(2):371–398.

CHAPTER 9

# Disaster Entrepreneurs, Emotions and Environmental Movements in France and in China

*Pierre Manoury*

As climate change and environmental degradation unfold over long temporal horizons, our capacity to recognize and respond to disaster is increasingly challenged. These slow-onset disasters lack clear and visible thresholds, making them harder to problematize and publicize. In this context, new forms of engagement emerge—shaped by ecological urgency but also by emotional uncertainty. At the intersection of environmental disruption and political constraint, a distinct figure is emerging within environmental movements: the *disaster entrepreneurs*. It refers to actors or groups of actors—such as organizations, activists, coordinators, or experts—who actively work to define and render visible slow-onset disasters. Building on the notion of moral entrepreneurs (Becker, 1963), disaster entrepreneurs denounce environmental slow-onset disasters by creating and enforcing situated rules. They embody the effects of the disaster, organize around it, and produce both emotions and interpretive frameworks that structure collective mobilization. These actors work to render slow-onset disasters legible, urgent, and actionable—within contexts that often resist such recognition. This chapter explores how disaster entrepreneurs mobilize emotions, construct frameworks of interpretation, and navigate public arenas to make sense of diffuse, delayed, and often invisibilized disasters.

Their roles are shaped by situated forms of suspicion which refers to a context-dependent emotional response to slow, uncertain forms of disruption. It signals a form of collective unease rooted in the latency, ambiguity, or invisibility of disaster effects. In France, this takes the form of *eco-anxiety*, a feeling of misalignment tied to systemic degradation. In China, it is expressed through *wúlì* (无力), a form of helplessness towards political constraints. Both function as emotional alloys (Illouz, 2007) that reconfigure how environmental actors engage with uncertainty.

Drawing on a multi-sited ethnography of four environmental organizations: two in France and two in China, this chapter hypothesizes that environmental slow-onset disasters produce situated forms of suspicion—such as eco-anxiety in France and *wúlì* in China—within environmental movements. These situated emotions contribute to the emergence and role-shaping of disaster entrepreneurs in public arenas. In France, they adopt roles centered on

 | HTTPS://DOI.ORG/10.1163/9789004751804-011

care, preparation, and the (re)production of frameworks. In China, they operate within a regime of constrained autonomy, enacting roles of enforcement, negotiation, and *societalization*. Taken together, these roles offer a situated lens through which to apprehend how slow-onset disasters shape both emotions and environmental mobilization across distinct political contexts.

## Methodology

This research is based on a multi-sited ethnography carried out within four environmental organizations—two in France and two in China. To preserve the anonymity of participants and institutional contexts, the groups are referred to here by pseudonyms: X and Y (France), and C and G (China).

In France, group Y is the local chapter of an international environmental NGO, institutionally anchored and vertically structured. Its activities focus on public direct action, outreach, thematic expertise, and policy advocacy framed through established institutional channels. In contrast, group X operates as horizontally organized grassroots collective grounded in regenerative culture. It emphasizes direct action, collective experimentation, and the cultivation of alternative modes of organization and care.

In China, both organizations studied differ significantly in structure and scope. Group C is a nationally prominent NGO composed of specialized departments and oriented toward institutional partnerships. Group G, by contrast, functions as the regional office of an international NGO, with a centralized internal hierarchy and transnationally defined strategic objectives. Despite these structural differences, both Chinese organizations prioritize expert knowledge, strategic communication, and the technical framing of environmental issues.

The empirical material collected includes participant observation in meetings and actions, semi-structured interviews with activists and staff members, internal documents (strategy papers, guidelines, online communication), and fieldnotes. Attention was paid to how emotions were expressed, regulated, or silenced within each organizational setting.

This comparative configuration—juxtaposing vertical and horizontal, national and transnational, centralized and self-organized structures—offers a situated analytical framework to investigate how slow-onset disasters reshape emotional dynamics within environmental movements, and how these emotions—particularly situated forms of suspicion—contribute to the emergence and differentiation of disaster entrepreneurs across distinct political and organizational contexts.

## 1 Slow-Onset Disasters, Situated Suspicion, and the Roles of Disaster Entrepreneurs

Long focused on sudden-onset events, disaster sociology has progressively expanded its analytical scope to encompass more diffuse forms of collective disruption—at the risk, some authors argue, of losing the very meaning of the word "disaster" (Giry, 2023). Since the 2000s, part of sociological literature has turned into elusive, diffuse, or chronically embedded forms of disasters that escape the classical grammar of the event. These include invisible pollution, ecosystem degradation, industrial contamination, and climate change. Such phenomena have led to the development of the concept of *slow-onset disaster*. In contrast to rapid disasters, these events lack both a clear moment of occurrence and an identifiable alert threshold. Their stretched temporality, structural embeddedness, and spatial dispersion render their qualification unstable, raising unprecedented issues of recognition, mobilization, and politicization.

Several authors argue that these disasters result less from rupture than from an accumulation of discrete disturbances, whose visibility depends on specific social dispositives: public inquiries, activist mobilization, and scientific controversies (Fiske, Marino, 2020). The event does not preexist these processes, it is constituted through the articulation of weak signals, the narration of fragmentary situations, and the construction of interpretive continuity (Gilbert, 1992). From this perspective, the slow disaster is a catastrophe without an event—grounded in a diffuse collective ordeal that is difficult to name, and whose politicization is constantly threatened by invisibilization.

This latency complicates collective mobilization. While disasters can lead various groups to construct political economies of judgment—producing regimes of consent, grievance, indignation, or institutional distancing (Roulleau-Berger, 2016)—they are also shaped by situated emotions. These emotions sustain forms of engagement led by intermediary actors who seek to make visible disturbances that have not yet been named. The emergence of slow-onset disasters as collective objects depends on active work of publicization. It is through this lens that the figure of the *disaster entrepreneur* takes shape.

This notion refers to actors and groups of actors who seek to enforce a moral rule of recognition of a disaster by defining certain behaviors as deviant, securing their social recognition as such, and ensuring the sanction of those who breach them (Becker, 1963). In the case of slow disasters, their role is even more central given the temporal and spatial dispersion of effects, which renders any clear designation of the event uncertain. The disaster entrepreneur produces *suspicion* in the face of the uncertainty of these disasters, often through varied careers marked by personal exposure and constrained engagement. Derived from *suspicere* ("to look up from below"), suspicion can be defined as a corpus of expressions of uncertainty within interactions—allowing us to grasp the hidden nature of situated regions of meaning and reality (Goffman, 1974). The use of this notion will help us to grasp the situated emotions produced by the actors in France and in China.

Since the 2016 Foreign NGO Law and the post-Covid era, the scope of action for social organizations in China has narrowed. Organizational survival has become a priority in a system marked by limited operational autonomy (Teets, 2013) and strategic ignorance logics (Hsu & Hasmath, 2016), where authorities tolerate the initiatives of environmental social organizations (*huanbao shehui zuzhi* 环保社会组织) as long as they do not challenge political balances. These dynamics are part of a consultative authoritarianism (Gao, Xia, 2016), in which civil participation is confined to acceptable registers. In this context, activists and employees of social organizations develop weakened forms of suspicion within constrained autonomy, without direct confrontation. Participation takes the form of contingent actions of reduced exposure (physical or digital signaling, occasional dissemination, reliance on consensual figures), constituting the role of disasters entrepreneurs who aim to avoid frame break while circumventing censorship thresholds.

In France, by contrast, slow-onset disasters unfold in more confrontational, open, and institutionalized public arenas (Denis, 2011). Disasters entrepreneurs emerge in varied roles: some engage in vertical, institutional structures, generating weakened suspicion and acting as rule enforcers; others operate in horizontal structures and therapeutic communities, producing intensified forms of suspicion, and taking on roles of care, preparation, or survival. The intensification of physical, emotional, and legal exposure (e.g. increased arrests, surveillance, or psychological toll) redefines their role: the disasters entrepreneur no longer only warns or raises awareness but acts to prepare for, or survive within, a present that is "unexpected, undesirable, unimaginable, unmanageable (Boin, 't Hart, 2007), and indefinable" (Giry, 2023).

## 2 Eco-anxiety, Wúlì and Mobilizations

In the Anthropocene era, eco-anxiety—understood as a kind of "homesickness without exile" (Morizot, 2019)—is emerging among certain activists in France. It acts as a weak signal of a broader collective and ecological disturbance, embedded in the semantics of risk. This disturbance takes the form of emotional alloys (Illouz, 2007), shaped by an urgent sense of fear, systemic ecological causes, and the appearance of new kinds of affliction (Laval, Pégon, 2024). The slow onset disaster triggers a sensitive, normative, and affective disturbance that reshapes the actors' representations and behaviors, provoking a form of affective misalignment (Latour & Schultz, 2022).

Eco-anxiety, seen as an aggregate of protean emotions, transforms both individual and collective interactions among those affected. These symptoms stem from a situated form of suspicion, fed by the uncertainty tied to the slow unfolding of the disaster. Yet this form does not capture the full spectrum of emotions among environmental activists in France. Some, even prior to their engagement, reject eco-anxiety, which they associate with passive uncertainty. They prefer terms such as eco-anger,[1] sadness, or nostalgia.

Within group X, activists experience an intensified form of this situated suspicion and root their commitment in an alternative ideological narrative that shifts the burden of ecological disturbance from the individual onto the system and its moral standards. The organizational frameworks in place support processes of emotional release and the reclaiming of anger (Flam, 2005), enabling the production of subversive counter-emotions. Through exposure, engagement, and repression, some activists undergo a reconfiguration of their emotional repertoire. They speak of anger, nostalgia, sadness, satisfaction, or even joy—but sometimes reject eco-anxiety, seen as an obstacle or even a form of "*self-sabotage*." One activist from group X expressed it this way:

> I used to feel anxious, like: what's going to happen, how will it go? I don't ask myself those questions anymore. I just go. We go. Then it's on. I really run on anger. It's a source of energy that lets me go all-in and avoid the self-sabotage of "can I really do this, is it a good idea ..." […] There's sadness, about what we're losing—the lives, the species, the ecosystems—but ... we deal with it however we can. The way we manage is by trying to find moments of beauty. And I think there's no shame in putting joy inside the sadness. […] We need to cultivate joy within sadness, joy within anger. I'm not anxious anymore. Sometimes I feel sorrow, I feel nostalgic—for when I was a kid and didn't worry about all this. I feel nostalgic for the life that's vanished from the garden. The more you act, the more resistance you meet, the less room there is for anxiety, because you're grounded in the present. There's no longer this projection of "what's going to happen." I associate anxiety a lot with being tossed around by events […] I'm in control of myself, and in those moments, I can feel sadness, anger, even

1 Interview with an activist from X group, March 2025.

> satisfaction, but I don't feel that anxiety I link to uncertainty, to not having chosen where you are—to a kind of passivity, of waiting.[2]

The interview given by this activist from group X highlights the sequential transformation of emotions. The eco-anxiety felt before joining X was reshaped and ultimately rejected, as it came to be associated with a form of passive waiting and uncertainty. This rejection is part of an alternative convention framework, shaped through a process of self-exposure and the redirection of emotional disturbance into action. Anger emerges as a central operator in this emotional work—an organizing energy that enables one to "stay in the present moment" and avoid anxious retreat. The activist engages in emotional management where emotions are transformed in their duration, direction, and intensity: nostalgia, sadness, or grief no longer obstruct commitment, but are instead made compatible with a daily practice of resistance and care.

The capacity to name loss, to generate and manage emotions in response to the degradation of the living world, to "*cultivate joy within anger*," sketches the contours of a new emotional relationship to the slow onset disaster. Far from a linear narrative of disillusionment or despair, which other activists may experience, this activist construct action-oriented competencies from a shared disturbance, whose handling is grounded in the (re)production of care, and prevention. The testimony thus becomes a valuable indicator of how activists feel and act within uncertainty—replacing anxiety with a situated form of control, articulated through practices of emotional subversion.

In the contemporary Chinese context, understanding environmental mobilizations requires close attention to the dynamics between action competencies, the feeling of helplessness, and political control mechanisms—a constellation that can be framed through the notion of constrained autonomy. The emotional response to this configuration finds expression in *wúlì* which refers to a subjective experience of being unable to directly act upon social, political, or institutional structures. It manifests as a form of individual resignation in the face of slow-onset disaster. This posture has developed in response to the accumulation of large-scale pollution episodes that have profoundly affected the Chinese population, particularly in 2013. That year, China faced highly publicized events including prolonged periods of extreme air pollution in Beijing and other cities, rivers contaminated by pig carcasses, and cadmium-laced rice in the south. These public opinion events (*yuqing shijian* 輿情事件) are defined as a sequence in which a local situation—often perceived as unjust or

2 Interview, activist from X group, March 2025.

alarming—gains national or translocal visibility through digital networks, triggering a social or institutional reaction. These events emerge from a process of publicization led by actors who select the facts, channels of dissemination, and discursive forms that allow locally unacceptable situations to be rendered visible (Fu, 2017; Yang, 2009).

Confronted with these risks, individuals were left to choose between "*staying indoors to avoid the effects of pollution or going out anyway.*"[3] This stance appears to reflect a regime of consent regarding the authorities' response to the slow-onset disaster—one over which individuals have no direct leverage. Such behavior may be interpreted as a posture of *wúlì*: an accepted helplessness, a way of "*making do*" with an uncertain environment, as revealed by this action with a group working with group C and the narratives of its participants.

> A waste collection activity was organized by a volunteer group in the mountains on the outskirts of Beijing, gathering around a hundred participants. The event, accessible via an online form including an environmental multiple-choice quiz and a registration fee, combined hiking with the optional opportunity to pick up litter. On site, volunteers could request gloves, bags, or grabbers to participate. The organization relied on coordinators responsible for both group safety and the management of collected waste. Several participants came from diverse backgrounds—lawyers, former members of other NGOs—and shared their attachment to nature, frustration with the constant reappearance of litter, and a desire to raise awareness without confrontation. The act of collecting waste was seen as a form of silent sensitization, driven by a love for the landscape and the need to preserve its beauty, despite a shared awareness of the structural limits of such actions.[4]

During this observational sequence, *wúlì* becomes visible through the entanglement of concrete engagement and internalized helplessness. This mode of action, though embodied, is grounded in an economy of restraint: it is not about confronting polluters or demanding systemic transformation, but rather about "keeping one's head down" and "continuing to pick up" in the hope of a slow shift in practices. This feeling of helplessness is more explicitly expressed in the volunteers' narratives. One participant explains that pollution has been pushed westward, out of the immediate visual field of urban Beijing residents,

3 Fieldnotes, informal interview, employee of C group, 7 June 2025.

4 Fieldnotes, C group activity, May 2024.

making the urgency invisible. For him, people's priority remains survival in a high-pressure society, relegating environmental issues to a secondary concern.

Another volunteer, shaped by her experience in Green River (*lüse jianghe* 绿色江河)—another environmental NGO—describes the frustration of seeing garbage reappear the very next day, and the inefficacy of isolated gestures. For these actors, helplessness lies not in inaction, but in the disconnect between concrete acts and systemic outcomes. Suspicion no longer concerns the truth of the facts, but rather their uptake by individuals. The act of collecting trash thus becomes a way of inhabiting the disaster.

Eco-anxiety and *wúlì* emerge as two situated forms of suspicion provoked by the diffuse experience of slow-onset disasters. Although rooted in different contexts, these emotional alloys reflect a shared tension: a sensitive disturbance born from the impossibility of fully acting on systemic disasters—imperceptible in their immediacy yet deeply structuring in their effects. In France, eco-anxiety functions as an indicator of a generalized affective misalignment, leading some activists to invest in public arenas and (re)produce frameworks that favor emotional expression, recognition, and transformation into tools for mobilization.

Conversely, in the Chinese context, *wúlì* operates as an internalized experience of helplessness—not toward the emotion itself, but toward its expression and potential political consequences. It does not represent a temporary misalignment but rather a durable adjustment to structural uncertainty, shaped by repeated pollution episodes yet tempered by the state-led publicization of discourses such as "*ecological civilization*" (*shengtai wenming* 生态文明).[5] *Wúlì* is inscribed within a regime of constrained autonomy, where action remains possible only insofar as it does not breach institutionalized conventions. It reflects a regime of consent to the impossibility of direct intervention and generates situated action competencies.

These two forms of suspicion are therefore not merely individual emotional responses, but indicators of how collectives perceive, organize, and address the effects of slow-onset disasters. While some French activists reproduce emotions within frameworks explicitly geared toward the politicization of disturbance, Chinese actors develop indirect, contingent, often silent responses that are nonetheless intentional.

---

5 The concept of "ecological civilization" (*shēngtài wénmíng* 生态文明) was first introduced in China's 17th Party Congress in 2007 and officially enshrined in the Constitution of the People's Republic in 2018. It presents a governance paradigm combining economic growth, ecological restoration, and green development, rooted in traditional ideas of harmony between humans and nature (e.g., Daoism).

## 3 (Re)Produce, Care and Prepare in France

The ways in which emotions are produced and managed within activist groups—shaped by situated forms of suspicion—foster specific roles for disaster entrepreneurs. The example of local chapter X shows how the implementation of practices through the SOS (self-organizing system), within the framework of regenerative culture, foster emotional expression and can serve as a temporary mechanism to mitigate emotional misconduct related to eco-anxiety. The SOS in X group is characterized by a horizontal, inclusive, and locally rooted structure, which values autonomy, decentralization, and the reduction of power concentration, while encouraging collective reflection and learning. This model stands in contrast with external frameworks perceived as vertical, exclusive, and centralized. Each local chapter establishes its own frameworks, produced and enforced by the militants present either physically or digitally. In group X, disaster entrepreneurs are the activists who actively participate in setting these frameworks, often taking facilitation roles in meetings and informal coordination tasks.

Another key pillar of the movement lies in regenerative culture, defined by the triad "*self-care, people-care, planet-care*" (Westwell, Bunting, 2020). This culture, closely tied to the SOS, goes beyond individual well-being to emphasize collective responsibility towards human communities and the environment. Based on this foundation, X's regenerative culture promotes the creation of "*non-immediate interdependence and reciprocity*" (Brugère, 2021), recognizing both individual vulnerability and dignity within the movement. Furthermore, this framework actively seeks to prevent violence—whether physical or symbolic—and challenges certain external conventions perceived as oppressive, as they require either active or passive emotional management. In this sense, these disaster entrepreneurs take on a role of care by designing and maintaining emotional and organizational spaces that make this reciprocity possible. This role lies in embodying the values of regenerative culture, and also in ensuring their continuity through practices and collective rituals.

During meetings, various practices are implemented by activists to embody this role and to produce and maintain new shared and explicit frameworks. During check-ins and check-outs, activists are invited—but not obliged—to state their pseudonym, pronouns, and "*inner weather*" (inviting each participant to briefly express their current emotional state). This ritual helps relieve the emotional work of activists by allowing them to express the emotions they are truly experiencing, rather than those they are expected to feel. During meetings, it is considered "ok"—a word frequently used to mark emotional validation—to cry, be angry, afraid, anxious, or joyful. This care role also appears during meetings and actions through the organization of "*artivism*" workshops (a portmanteau of art and activism), allowing rebels to express themselves through dance, cinema, singing, painting ... These activities form care techniques aimed at preventing the effects of eco-anxiety, overcommitment, and repression.

> One of the groups was tasked with building cardboard cars and a cardboard planet as part of a trial involving activists who had carried out an action on October 2022. They had held up signs over luxury cars, splashing them with paint, and some had glued their hands to the vehicles. The atmosphere during this activity was joyful, creating a space for emotional expression.[6]

Some militants even adopt care techniques outside the activist context (vegetarianism, disconnection, recycling, repairing ...), integrating their commitment into their lifestyle, marked by a concern for individual and collective well-being and sustainability. In this sense, care goes beyond the private sphere to become a central component of the disaster entrepreneur's role as rule enforcer, reaffirming the sensitive interdependence between self-care and care for the planet that defines regenerative culture.

This role is institutionalized through a mandate conferred by vote to two group members for a six-month term, during which they become responsible for regenerative culture. On the group's digital platform, the mandate description for this role is the longest and most imprecise in terms of specific tasks. For instance, "*Fighting against all forms of physical or verbal violence, discrimination, and oppression*" is not a task *per se*, but rather the formalization of a rule. In practice, the designated activist rarely intervenes and are seldom solicited. Nonetheless, they act as enforcers of the prescribed rules outlined in this mandate within the local chapter.

Yet, the emotional and physical exposure of activists from group X, within a French context that is becoming increasingly less democratic, generates emotional alloys that intensify suspicion and shift the role of these disaster entrepreneurs toward one of preparing for societal collapse caused by the slow onset disaster. In doing so, they take on the role of preparing for physical survival and for the moral and emotional dislocations that collapse implies. Preparation becomes a way to re-anchor meaning in an uncertain environment.

6 Fieldnotes, weekly meeting, Group X, 9 October 2023.

During the 2024 Olympic and Paralympic Games in Paris, X activists were subjected to four waves of police raids, along with numerous detentions. Several activists interviewed after these operations appeared deeply affected. This distress manifested in sleep disorders, crying, paranoia … One activist, after experiencing a raid in the summer of 2024, systematically searched his apartment for potential hidden listening devices, leading to stress, insomnia, and anxiety. The fear and anger—already heightened prior to these events toward law enforcement—were further exacerbated by these repressive actions. This gradual shift in context gave rise to ongoing situations of uncertainty, rendering the exceptional ordinary, diminishing the capacity for action, and fueling anxiety about an unlivable future, as expressed by this X activist speaking of "disenchantment":

> Since I allowed myself to say it out loud, I'm sad that this action (Olympic games action) didn't work out, I'm sad that we got s****** over so hard by the police, I'm sad that the RN (Rassemblement National) is scoring so high […], I feel even less in the mood to take action […]. We were starting to reach a point where I needed to take a break from activism, because I … disenchantment. Today, I've found a way to live with it, but I don't have the answer. And I'm not necessarily doing better, I just live with it.[7]

The increasing repression by law enforcement in France, the inability of movement X to meet its goal of carbon neutrality by 2025, and the growing sense of powerlessness in the face of the rapidly worsening global slow onset disaster have led to new sequences in activists' careers. Within group X, many have experienced partial or total disengagement, or have re-engaged in more radical movements. Activists are thus redefining their relationship to the collective trouble and gradually adopting a new role as disaster entrepreneurs. From raising awareness about the slow onset disaster, they shift toward preparation and survival in anticipation of societal collapse due to this disaster.

> For me, it's a logic of survival, it's not a choice […] Collapse is a society that can no longer guarantee all basic services simultaneously through democratic means. Either it compromises, imposes temporarily, or enforces through coercion or undemocratic means. That's what a collapsing society looks like, and we're already seeing early signs of it. […] We're in a pre-collapse situation. […] With the breakdown of ecological conditions,

7 Interview with an activist, X group, March 2025.

> with biodiversity loss, everything else collapses too. Environmental struggle is more important because it affects absolutely everyone.[8]

This reconfiguration of roles and commitments within group X contrasts with the relative stability observed in other organizations. While some actors radicalize or disengage in response to repression and failure, others remain anchored in more institutionalized and enduring structures. Within group Y, the figure of the disaster entrepreneur takes on the role of reproducing established rules. These activists mainly ensure continuity within the institutionalized frameworks of the organization, acting primarily as rule enforcers. These disaster entrepreneurs maintain their focus on mitigating climate change without adopting logics of collapse preparation. Rooted in an older, more institutionalized organizational structure, activists retain stable modes of action and discourse, with fewer sequences of disengagement or radical re-engagement. Their role as rule enforcers established by the central organization contrasts with the dual role of rule creator and enforcer observed among group X activists.

Moreover, eco-anxiety appears to be largely absent or significantly diminished within this group, affecting how care is integrated into their organizational life. For example, while entrance rounds and non-violent communication practices have been introduced during meetings, these new rules are seldom enforced or consistently followed. Here, the disaster entrepreneurs primarily reproduce institutionalized frameworks. They ensure continuity rather than rupture, prioritizing procedural legitimacy over emotional innovation. This form of reproduction stabilizes activism but limits its transformative potential in the face of evolving disruptions.

> During a monthly meeting in the observed local chapter (Y), an attempt to introduce non-violent communication (NVC) practices was initiated by a group of young female activists. While these proposals initially elicited polite receptiveness, they were quickly sidelined in practice as the meeting progressed. The activist capital of older generations seemed to trigger a spontaneous alignment of interactional frameworks, drawing exchanges back toward more direct interaction, often loaded with implicit hierarchies. Although explicitly valued in discourse, the gestures of NVC appeared to be neutralized by the enduring influence of earlier activist socializations according to rules vertically and historically

8 Interview with an activist, X group, March 2025.

> prescribed by the organization headquarters, perceived as legitimate and deeply internalized.[9]

This limited uptake of care techniques suggested by younger activists within group Y does not appear to stem from explicit rejection, but rather from a process of invisibilization linked to the role of rule enforcers played by the group's members. The rules produced by the organization's headquarters—verticalized and directive—do not allow for the horizontal creation of new frameworks within group Y as in X group.

This structural separation between action and emotions does not neutralize internal tensions but rather contributes to their displacement. Attempts—often driven by younger generations—to introduce tools such as non-violent communication or check-in rituals are frequently absorbed without being truly institutionalized. They become weak signals of a desire for framework recomposition but struggle to take root against the strength of stabilized ones. In this sense, group Y embodies a weakened form of suspicion, where emotions are neither denied nor fully embraced, but managed through omission, enacting a role of reproduction and maintenance rather than one of creation and care.

Whereas the situated forms of suspicion observed in the French case led to a redefinition of the role of disaster entrepreneurs through the (re)production of frameworks, practices of care, and preparation, this process takes a different shape in contexts where public arenas are more tightly constrained by the state. In China, emotional alloy such as *wúlì* reorient the role of these entrepreneurs toward enforcing, negotiating, and *societalizing*. They aren't defined by the capacity to establish new frameworks, but by their ability to produce practical competencies within the context of constrained autonomy.

## 4 Enforce, Negotiate and Societalize in China

In his study on the structuring of public arenas in China, Shawn Shieh identifies three modes that describe this process. First, regulations, which is imposed vertically by the state on social organizations. Then, the mode of negotiation, characterized by voluntary interactions that nonetheless remain asymmetrical and vertical between social organizations and the state. Finally, the third is *societalization*, where actors bypass direct confrontation with the state by mobilizing social resources involving horizontal and symmetrical exchanges

9 Fieldnotes, meeting, 14 September 2024.

with other organizations or the broader population (Shieh, 2016). In this chapter, the latter two modes are approached not as institutional dynamics but as roles performed by disaster entrepreneurs, along with a third role of enforcement, which mirrors the vertical mode of regulation by applying rules defined by the state. These three roles provide a useful framework to analyze how disaster entrepreneurs operate within constrained political environments, with the emotional alloy of *wúlì* shaping them and enabling transitions between them.

*Wúlì* manifests when actors perceive that the consequences of their actions—even when these receive media or institutional attention—can be reframed or fabricated by local or central authorities, thereby redirecting the process of publicization reinforcing a sense of helplessness. Such manifestations of *wúlì* are often intensified in public opinion events, where diffuse grievances crystallize into nationally visible controversies. This intensification of *wúlì* can, trigger transitions between roles, as disaster entrepreneurs—often organizations rather than individuals in the Chinese context—adapt their strategies to navigate shifting constraints.

One example of action carried out by organization C, as part of one of these events in Hebei province, highlights the importance of public attention in the intensification of *wúlì* but also in the expression of the roles. In this province, the local government planned a construction project on the *Huaihe River* (淮河), which would have destroyed the habitat of a particular species of migratory bird, the sand martins. After being informed by volunteers, C sent an official letter to the local authorities requesting that the project be halted and that the organization be invited to conduct an environmental impact assessment. The authorities subsequently invited C to assess the situation on-site. In this sequence, organization C acted as a disaster entrepreneur by initially adopting a role of negotiation with local authorities. However, upon arrival, its representatives were denied access to key environmental sites, restricting the organization's ability to carry out its investigative role and highlighting the limits of negotiated engagement within a constrained institutional environment.

In response to this obstruction, organization C mobilized various social media platforms to publicize the case and generate public attention, thereby triggering a public opinion event. This shift marked a transition from a mode of negotiation to one of *societalization*: disaster entrepreneurs within the organization changed their role by relying on mediated publicization and social resonance, rather than negotiation with local government.

> The whole society started paying attention to it, which puts a lot of pressure on the local government. Eventually, they agreed not to remove the

> sand dunes. They fenced off the area and asked people not to enter. As a result, the birds were able to live and breed there again.[10]

The local government eventually yielded and chose to authorize C to conduct a full investigation, to preserve the sand dunes where several sand martin nests had been identified, and to erect barriers preventing public access to the area, in order to avoid the destruction of the environment and prevent potential disturbances to the species present. This shift back to a role of negotiation occurs once the role of *societalization* has produced sufficient public pressure to prompt a response from local authorities.

> Indeed, these public opinion events, which can be either positive or negative, have a significant impact on the organization's development, its funding, and the political and popular pressures it receives. However, the organization pointed out that, in another sense, this phenomenon could serve its interests by channeling public attention toward environmental abuses, in order to sanction—or prompt the sanctioning of—certain officials.[11]

However, the configuration of these roles—shaped by *wúlì*, a situated form of suspicion arising from the tension between action competencies and structural constraints—unfolds within a context that is increasingly restrictive for social organizations. The restrictions include formal constraints such as the 2016 Foreign NGO Law,[12] informal censorship thresholds, and unwritten rules regarding acceptable forms of engagement. The central state anchors the transition from a system of overt sanctioning to a model of tacit sanctioning in China's public arenas, aiming to confine social organizations to the role of intermediary actors (Hsu & Hasmath, 2016).

Within this logic, disaster entrepreneurs navigate and uphold the tacit limits of expression and intervention tolerated under authoritarian rule. They

---

10 Interview, employee in C group, 2022.

11 Fieldnotes, meeting, 6 June 2024.

12 The formal title of the 2016 Foreign NGO Law is "Law of the People's Republic of China on the Administration of Activities of Overseas Non-Governmental Organizations within Mainland China" (中华人民共和国境外非政府组织境内活动管理法), adopted by the National People's Congress on April 28, 2016 and implemented on January 1, 2017. It mandates that foreign NGOs must register with the Ministry of Public Security or affiliate through a Chinese sponsor, restricts their engagement in political or religious activities, and grants broad oversight powers to police. Violations—including operating without registration—can result in warnings, suspension, confiscation of assets, or expulsion.

internalize the rules and manage them, ensuring that environmental claims remain legible without triggering repression. For instance, Group G, being a regional office of an international NGO, has not yet secured the backing of a governmental sponsor required under the 2016 Foreign NGO Law and is therefore unregistered and illegal. However, it operates within a framework of strategic ignorance and is tacitly tolerated, if it systematically cooperates with organizations supervised by state-appointed representatives.

> For G group, because we don't have the registration, we don't have one governmental department to approve our plans. So every project we want to do, we first need to find a Chinese partner, like a university. Then we work together on a work plan, which needs to be approved by the management office of our Chinese partner. It can be under the Ministry of Education, or sometimes another government department. So we try to keep our work plan general—we don't put too many details. After approval, we send the plan to the police who needs to approve all the work plan. [...] We usually don't have our project in just one city or one province. We write to as many provinces as we can—maybe ten or more—so as if we get rejected like in Beijing, we try another place. We have to do our best to squeeze into space. If we stop, the space will just keep getting smaller and smaller. I think in China, this kind of space is like a rubber band—a xiàngpíjīn (橡皮筋). The rubber band is around us and becomes smaller and smaller. But if we try to push in some area and if we feel a little bit of pain we come back. Then maybe we can push somewhere else. So as long as the rubber band hasn't broken, and we're not punished by law and our office isn't closed, we always have to be flexible and try other possibilities.[13]

This testimony provides an ethnographic illustration of how disaster entrepreneurs operate within the shifting contours of tacit sanctioning. It shows how *wúlì*, a structuring emotional alloy, orients the organization's role in navigating ambiguous legal status. Rather than contesting restrictions, the group produces pragmatic and adaptive action competencies—described metaphorically as navigating a rubber band—which consist of testing boundaries, retreating under pressure, and redirecting initiatives elsewhere. Group G thus takes on the role of *societalization* by activating social resources—such as academic or semi-official partnerships—to bypass direct confrontation with the

13 Interview, employee in G group, June 2024.

state, while preserving its capacity to render the slow-onset disaster visible through situated action.

This situated navigation does not produce new rules but rather operates within existing ones. Accordingly, the work of these disaster entrepreneurs is not that of rule creators but rather of rule enforcers within public arenas. The study of these practices reveals how slow-onset disasters can be requalified as public problems in authoritarian contexts—not despite, but through political constraints. In this context, citizens are embedded in a condition of constrained autonomy: they act within and through the limits imposed by political power.

There are thus several continuities in the effects produced by slow disasters on environmental movements in both France and China. In both contexts, these diffuse disasters trigger a sensitive disturbance that unsettles habitual perception frameworks, generating complex emotions such as *wúlì* or eco-anxiety. These emotions are not merely individual experiences but reflect an affective misalignment in response to perceived helplessness and the worsening of ecological conditions. They contribute to the reconfiguration of action competencies and redefine the roles of disaster entrepreneurs. In both cases, publicization processes, though situated, appear as structuring sequences where suspicion becomes operative in making the disaster visible. The disaster entrepreneurs in China thus oscillate between enforcement, negotiation and *societalization*, producing competencies to seize opportunities for action while managing the emotional costs of institutional ambiguity.

However, these forms of suspicion unfold within deeply situated socio-political regimes, which distinctly shape both the capacity for action and the recognition of disruption. In France, eco-anxiety is collectively translated through dispositive such as the SOS or regenerative culture and becomes a resource for producing alternative frameworks. In contrast, in China, *wúlì* is an internalized form of dissonance, rooted in a regime of constrained autonomy, where emotions do not lead to direct politicization.

## 5 Conclusion

The study of emotions among environmental movement actors facing slow-onset disasters in France and China reveals situated forms of suspicion that deeply reconfigure action competencies and the roles of disaster entrepreneurs. These actors embody pivotal figures in the (re)production, circulation, and preservation of emotions in a world where disaster is lived more than seen. The case of group X illustrates the intensification of a suspicion experienced both as an emotional alloy (sadness, eco-anger, nostalgia) and as an organizational mode, structured through care techniques. The case of group Y illustrates a weakened form of suspicion, experienced less as an emotional alloy than as framework inertia, structured through the enforcement and reproduction of institutional framework rather than the creation of innovative ones or the assumption of care-oriented roles. In China, *wúlì* does not signify the end of engagement, but rather its displacement toward new roles, shaped by logics of enforcement, *societalization* and negotiation in a political environment where publicization itself may become a risk.

Thus, disaster entrepreneurs—whether they care, negotiate, enforce, or *societalize*—emerge as emotional indicators of the slow onset disaster. They embody the tensions between exposure and withdrawal, between framework affirmation and circumvention, between anticipated collapse and habitable present. Their role is not merely to respond to disaster, but to actively recompose ways of living in the face of uncertainty.

## Bibliography


Becker, Howard S. (1985). *Outsiders: Studies in the Sociology of Deviance*. New York: Free Press.

Boin, Arjen, and 't Hart, Paul. (2007). "The Crisis Approach." In *Handbook of Disaster Research*, edited by Rodríguez, Havidán, Quarantelli, Enrico L., and Dynes, Russell R., 42–54. New York: Springer.

Brugère, F. (2021). *L'éthique du care*. Paris: Presses Universitaires de France.

Chartier, Denis. (2020). « Les organisations non-gouvernementales ». In *Dictionnaire critique de l'Anthropocène*. Paris: CNRS Éditions.

Fiske, S.J., & Marino, E. (2020). Slow-Onset Disaster: Climate Change and the Gaps between Knowledge, Policy, and Practice. In S.M. Hoffman & R.E. Barrios (Eds.), *Disaster Upon Disaster: Exploring the Gap Between Knowledge, Policy and Practice* (1st ed., Vol. 2, pp. 139–171).

Flam, H. (2005). Emotions' map: A research agenda. In *Emotions and social movements* (p. 22). London: Routledge.

Fu, Diana. (2017). "Disguised Collective Action in China." *Comparative Political Studies* 50 (4): 499–527.

Gao, Bingzhong, and Xia, Xunxiang. (2016). "The Emergence of an Autonomous Social Sector in China." In NGO Governance and Management in China, 29–45. New York: Routledge. (*Routledge Studies on China in Transition*).

Gilbert, Claude. (1992). *Le pouvoir en situation extrême: catastrophes et politique*. Paris: L'Harmattan.

Giry, Benoit. (2023). *Sociologie des catastrophes*. Paris: La Découverte.

Goffman, E. (1974). *Frame analysis: An essay on the organization of experience*. Cambridge, MA: Harvard University Press.

Hsu, Jennifer Y.J., and Hasmath, Reza. (2016). "Governing and Managing NGOs in China: An Introduction." In *NGO Governance and Management in China*, 1–9. New York: Routledge. (*Routledge Studies on China in Transition*).

Illouz, Eva. (2007). *Cold Intimacies: The Making of Emotional Capitalism*. Cambridge: Polity Press.

Jobin, Paul, Ming-sho Ho, and Hsiao-Hung M. Hsiao (eds). (2021). *Environmental Movements and Politics of the Asian Anthropocene*. Singapore: ISEAS Publishing.

Latour, Bruno, and Christophe Schultz. (2022). *Mémo sur la nouvelle classe écologique*. Paris: La Découverte.

Laval, Christian, and Pégon, Guillaume. (2024). *Souci écologique et santé mentale dans un monde troublé*. Toulouse: Érès. (*Études, recherches, actions en santé mentale en Europe*).

Morizot, Baptiste. (2019). Ce mal du pays sans exil. Les affects du mauvais temps qui vient. *Critique*, 860–861(1).

Roulleau-Berger, Laurence. (2016). "Life Exposed: Inequalities and Moral Economies in Post-Disaster Societies—China, Japan and Indonesia." In *Ecological Risks and Disasters: New Experiences in China and Europe*, edited by Li Peilin and Laurence Roulleau-Berger, 135–154. London and New York: Routledge (China Policy Series).

Shieh, Shawn. (2016). "Mapping the Dynamics of Civil Society in China: A Modal Analysis of Trends in the NGO Sector." In *NGO Governance and Management in China*, 45–69. New York: Routledge. (*Routledge Studies on China in Transition*).

Teets, Jessica C. (2013). "Let Many Civil Societies Bloom: The Rise of Consultative Authoritarianism in China." *The China Quarterly* 213: 19–38.

Westwell, E., & Bunting, J. (2020). The regenerative culture of Extinction Rebellion: self-care, people care, planet care. *Environmental Politics*, 29(3), 546–551.

Yan, Fei, and Mingxuan Liu. (2024). "事件社会学——从'结构—事件'到'关系—事件' [Event Sociology: From 'Structure–Event' to 'Relation–Event']." 社会学研究 (*Sociological Research*), no. 1: 204–225.

Yang, Guobin. (2009). *The Power of the Internet in China: Citizen Activism Online*. New York: Columbia University Press.

CHAPTER 10

# Disasters, Livelihood, and Indigenous Cosmology: the Case of the Na in Sichuan and Yunnan (China)

*Jin Chen*

Disasters are multifaceted phenomena that transcend mere physical events, encompassing complex interactions between humans, their environments, and socio-cultural factors (Hewitt, 1983). In recent years, the field of disaster studies has gained prominence within the social sciences as researchers increasingly recognize the importance of understanding disasters within broader social, cultural, and historical contexts (Oliver-Smith, 1996). This recognition stems from the realization that effective disaster management and mitigation strategies must address not only the physical impacts of disasters but also the social, economic, and cultural dimensions that shape human vulnerability and resilience (Cutter et al., 2008; Gaillard et al., 2017).

One of the key contributions of disaster studies to the social sciences lies in its interdisciplinary nature. By drawing on insights from disciplines such as anthropology, sociology, geography, political science, and environmental studies, disaster researchers are able to explore the intricate connections between human societies and their environments (Bankoff, 2004; Oliver-Smith and Hoffman, 2002). This interdisciplinary approach allows for a nuanced understanding of how social structures, power dynamics, cultural beliefs, and historical legacies influence vulnerability to disasters and shape responses to them (Blaikie et al., 2014; Wisner et al., 2012).

Furthermore, disaster studies play a crucial role in amplifying marginalized voices and perspectives, particularly those of indigenous communities and ethnic minorities (Enarson and Morrow, 1998; Mercer et al., 2010). Indigenous knowledge systems, passed down through generations, offer valuable insights into sustainable resource management, traditional land-use practices, and community resilience strategies (Berkes, 2008; Huntington, 2000). By centering indigenous perspectives in disaster research, scholars can uncover alternative approaches to disaster preparedness, response, and recovery that are rooted in local knowledge and cultural traditions (Cajete, 1999; Whyte et al., 2016).

In this article, I explore the concept of natural disasters through the lens of local knowledge and modes of living. My research, spanning from 2003 to

 | HTTPS://DOI.ORG/10.1163/9789004751804-012

2019 among the Na (纳) in Southwest China, suggests that their understanding of natural disasters is intimately linked to their unique cosmological beliefs and day-to-day livelihood. Grasping how indigenous populations perceive and interact with their ecosystems is critical. This involves, firstly, an examination of the evolution of disaster anthropology, with a particular focus on environmental issues in present-day China. Secondly, I delve into the Na's relationship with their surrounding ecosystems. Thirdly, I discuss the cosmological importance of two types of spiritual entities: mountain and water spirits. Prior to concluding, I consider the recent socio-economic transformations within the Na society and their implications for environmental sustainability.

## 1 Anthropology of Disaster and Contemporary China Studies

In elucidating the developmental trajectory of the anthropology of disaster, it is imperative to incorporate contemporary scholarship that reflects the evolving nature of the field. The foundational work of Julian Steward (1955) and Anthony Oliver-Smith (1979) remains pivotal, providing early insights into adaptive strategies and the social dimensions of disasters, respectively. Steward's theory of culture change and methodology of multilinear evolution provided a framework for understanding how societies adapt to their environments, a key aspect when considering responses to environmental hazards. Oliver-Smith's early work emphasized the social fabric's alteration and the ethical considerations for anthropologists working in disaster contexts. The theoretical shift towards a social constructionist paradigm is further enriched by the contributions of Anthony Giddens (1994) and the nuanced perspectives of Susanna Hoffman (2003). Giddens explored how modern societies perceive and manage risk, highlighting its reflexive character, while Hoffman's work, particularly in relation to Oliver-Smith, has shed light on the ways disasters are culturally framed and understood. Their examinations of the discursive construction of disasters underscore the ongoing relevance of this theoretical orientation in understanding the socio-cultural dimensions of calamities.

To encapsulate the contemporary interdisciplinary turn in the anthropology of disaster, one needs to draw from the scholarship of recognized figures such as Anthony Oliver-Smith (2019), whose recent work delves into the complexities of post-disaster recovery and the ethical dimensions of applied anthropology in disaster contexts. This interdisciplinary approach, drawing from sociology, geography, and environmental studies, is crucial for a holistic understanding. For instance, Kai Erikson's (1976) seminal study on the Buffalo Creek Flood provided profound insights into the long-term social and psychological trauma within a community, demonstrating the deep interplay between sociology and disaster studies by focusing on the "destruction of community" as a core impact of disasters. This aligns with the broader understanding that disasters are not just physical events but deeply social and psychological experiences. Furthermore, the work of W. Neil Adger (2006) on vulnerability has been influential in shaping how disaster studies approach the differential impacts on populations, emphasizing that vulnerability is a socially constructed and differentiated condition, often predating the disaster event itself and tied to factors like socio-economic status, governance, and access to resources. Such perspectives are critical in analyzing why some communities or individuals suffer more than others.

Moreover, the field has witnessed a growing engagement with environmental anthropology and climate change issues. Noteworthy contributions include the works of Michael R. Dove (2016), who explores the intersections of environmental change, vulnerability, and adaptation strategies within diverse socio-cultural contexts, often challenging dominant narratives about environmental knowledge and management. This engagement is particularly relevant as climate change is expected to increase the frequency and intensity of certain types of disasters. The critical role of amplifying marginalized voices, especially those of indigenous communities, is emphasized by scholars like Enarson and Morrow (1998) through a gendered lens, and Mercer et al. (2010) who advocate for participatory research methods to genuinely incorporate local perspectives in disaster risk reduction. This resonates strongly with the insights from Berkes (2008) on "Sacred Ecology," which details how indigenous knowledge systems often contain sophisticated understandings of ecological relationships and sustainable practices that are vital for resilience. Huntington (2000) further elaborates on methods for integrating traditional ecological knowledge with scientific approaches, which can lead to more effective and culturally appropriate disaster mitigation strategies. Cajete (1999) and Whyte et al. (2016) also contribute by centering indigenous science and community-led solutions, particularly in the context of environmental and climate justice, which are intrinsically linked to disaster vulnerability and response.

Anthropological research on disasters in China has gained significant momentum, particularly following the transformative 2008 Wenchuan earthquake. Leading Chinese scholars like Li Yongxiang have been instrumental in developing disaster anthropology as a distinct field, emphasizing that disasters transcend mere natural phenomena to become profound socio-cultural events shaped by their cultural construction (Li, 2012). This perspective aligns with international scholarship that views disasters as complex processes revealing intricate interactions between natural, biological, and socio-cultural factors.

Contemporary disaster anthropology in China encompasses diverse catastrophic events—earthquakes, floods, droughts, and epidemics such as SARS—examining the complete disaster lifecycle from initial risk assessment and vulnerability evaluation through coping strategies and relief management. This comprehensive approach provides crucial insights into how communities prepare for, experience, and recover from major disruptions.

The 2008 Wenchuan earthquake emerged as a pivotal moment for anthropological inquiry, generating extensive studies on post-disaster reconstruction, social support networks, community rebuilding processes, and psychological impacts on survivors. Researchers have explored how such catastrophic events compel affected populations to fundamentally re-evaluate their reality and life values, with particular attention to the healing power of collective storytelling in community recovery (Huang et al., 2014). The concepts of "cultural recovery" and "culturally sensitive reconstruction" have become central themes, especially regarding ethnic minority communities like the Qiang (羌) people, who suffered disproportionate impacts from the Wenchuan earthquake. Scholars have critically examined state-sponsored heritage tourism initiatives and "gratitude education" campaigns in post-disaster contexts, analyzing their complex effects on ethnic identity, cultural preservation, and underlying power dynamics. Qiaoyun Zhang's research exemplifies this focus, investigating the intersection of disaster tourism and heritage tourism among the Qiang community. Her collaborative work with Roberto Barrios explores the politics embedded in culturally sensitive reconstruction efforts (Zhang & Barrios, 2017). While these initiatives aim to promote economic recovery and demonstrate state capacity, critics argue they sometimes fetishize cultural elements while overlooking the genuine trauma experienced by affected populations (Le Mentec & Zhang, 2017). The transformation of traditional Qiang villages into tourism destinations illustrates these complex dynamics between cultural preservation and commercialization.

The historical and ecological dimensions of disasters in China offer rich terrain for anthropological investigation. Scholars like Tristram R. Kidder have uncovered ancient patterns of human-environment interaction, including large-scale levee construction along the Yellow River that demonstrates how human interventions have historically shaped flood disasters and environmental change (Kidder & Liu, 2017). These studies reveal that efforts to control the Yellow River through engineered structures, dating back nearly three millennia, sometimes paradoxically exacerbated flooding by causing sediment accumulation and elevating the riverbed above surrounding floodplains.

The study of social vulnerability represents another critical dimension of disaster anthropology in China. Rapid urbanization and development have, in many instances, increased social vulnerability by creating uneven living conditions and marginalizing certain groups, making them more susceptible to disaster impacts (Xu et al., 2024). Anthropological and geographical studies assess vulnerability through comprehensive examination of exposure, sensitivity, and adaptive capacity, highlighting how factors such as limited education, poor health, unstable income, and inadequate social security contribute to heightened vulnerability among specific populations.

The Chinese government's evolving approach to disaster management, including the establishment of relief funds and the increasingly complex role of NGOs and grassroots volunteerism, provides valuable opportunities for anthropological investigation into governance structures, social participation, and humanitarian practices in disaster contexts (Hirono & Rose, 2020). The ongoing tension between state narratives of progress and local experiences of disaster and recovery continues to generate dynamic research, revealing both convergences and divergences in how state power and cultural meanings shape disaster response and recovery processes.

The expanding body of research reflects growing efforts to integrate social sciences with natural sciences to better inform disaster prevention, governance, and reconstruction strategies in China. This interdisciplinary approach emphasizes the critical need for culturally sensitive and community-based methodologies that respect local knowledge while addressing broader systemic vulnerabilities (Button & Schuller, 2021). As China continues to face diverse disaster challenges, anthropological perspectives remain essential for understanding the complex social, cultural, and political dimensions of disaster experiences and recovery processes.

## 2 The Livelihood of the Na and Their Relationship with Disasters

The Na people, often referred to as Mosuo (摩梭) by the Han, represent a unique ethnic group in China's southwestern region. Their self-identifications, such as "nazi", "nahing", and "naru", embody their essence as "people" or "human beings". The Na, with an estimated population between 30,000 and 40,000, inhabit the scenic and undulating landscape of the Qing-Tibet plateau, nestled at the juncture of Sichuan and Yunnan provinces. This area, part of the expansive Hengduan mountain range, is characterized by its dramatic topography of craggy mountain ranges and major rivers including the Yangtze, Mekong, and Salween. The region's remarkable geography has been recognized by UNESCO as part of the "Three Parallel Rivers" World Heritage Site in 2003.

The Na's traditional economy revolves around agriculture and animal husbandry, cultivating crops such as maize, wheat, and hemp, and rearing a variety of livestock including cattle, sheep, yaks, and horses. Their economic activities extend beyond farming to include spinning, weaving, and trading. The Na's historical governance by feudal "Tusi (土司)" chiefs evolved drastically following the 1950s under the communist regime. Spiritually, they have been adherents of Tibetan Buddhism since the Yuan dynasty, and their language, part of the Tibeto-Burman group, is unique for its lack of a written script.

The Na's symbiotic relationship with their mountainous environment is evident in their daily practices and cultural traditions. Their agricultural methods are perfectly adapted to the local terrain, with crop cultivation and livestock rearing tailored to the region's ecological conditions. Wooden houses, constructed using timber from local forests, blend seamlessly into the landscape, reflecting the Na's deep understanding and respect for their natural surroundings. Herding practices, especially of yaks in high-altitude pastures, showcase their adeptness in utilizing diverse environmental zones. The collection of medicinal herbs and mushrooms is not just an economic activity but also a testament to their extensive knowledge of the local flora. Water plays a pivotal role in the Na's daily life. Springs and wells are crucial for providing clean drinking water, and the task of water collection often falls to the women of the community, signifying its importance in social and domestic spheres. The communal use of hot springs for bathing and socializing further illustrates the integral role of natural resources in their societal fabric.

The Na's agricultural practices and ritual traditions exhibit a rhythmic regularity, reflecting a harmonious balance between their livelihood and cultural identity. This regularity is not just a mere coincidence but a representation of their profound connection with the natural cycles and rhythms of their environment.

The Na's relationship with disasters is deeply embedded in their agricultural calendar and ecological practices. Their traditional ecological knowledge has evolved over centuries to cope with various environmental hazards that characterize the mountainous terrain of the Hengduan range. The dramatic topography, while providing diverse ecological niches for their subsistence activities, also exposes them to multiple disaster risks including landslides, floods, droughts, and extreme weather events.

The Na's disaster preparedness is reflected in their agricultural diversification strategies. By cultivating crops at different altitudes and maintaining a variety of livestock species, they create a buffer against localized disasters. For instance, when frost damages crops in lower valleys, highland barley and potatoes often survive. Similarly, their practice of storing grains in elevated wooden structures protects food supplies from flooding while maintaining ventilation to prevent mold during the humid monsoon season.

TABLE 10.1 Agricultural and ritual activities of the Na around the year

| Month | Ordinary activities | Ritual activities |
|---|---|---|
| January | Celebrating the New Year, preparations for working in fields ("krushiloshi") | Offering to ancestors ("dhaladhu") and to the spirits of mountains ("dzabu"), etc. |
| February | Sowing ("du") potatoes and oats | "nadZido" (buddhistic ritual to venerate the spirits of springs) |
| March | Sowing maize ("kradzéshua"), hemp ("sa") and beans | |
| April | First weeding ("dza") of maize, picking edible wild herbs | Venerations of the spirits of springs ("gubu") and of mountains ("dzabu") |
| May | Harvest of barley and honey, second weeding of maize | "NuohlimiZi" (seasonal offerings to ancestors) |
| June | Harvest of potatoes | |
| July | Sowing turnip, harvest of hemp ("sami") | |
| August | Harvest of beans and buckwheat | "yibu" (seasonal offerings to ancestors) |
| September | Harvest of maize, sowing barley, oats and Qingkre (local barley) | |
| October | Harvest of turnip and honey | "busina" (great offerings to ancestors), venerations of the spirits of mountains ("dzabu") |
| November | Lumbering wood, digging manure | "mukrabu" (expulsions of malefic spirits), "didha" (seasonal offerings to ancestors) |
| December | Cleaning residences | |

Their traditional housing architecture demonstrates sophisticated disaster mitigation techniques. The wooden structures, built with interlocking joints rather than nails, provide flexibility during earthquakes—a common occurrence in this seismically active region. The raised foundations protect against flooding and moisture, while the sloped roofs efficiently shed heavy snow and rain. The communal nature of construction ensures rapid rebuilding after disasters, with entire communities mobilizing to help affected households.

Water management practices among the Na reveal another dimension of disaster preparedness. Their intricate knowledge of seasonal water patterns allows them to predict and prepare for both floods and droughts. Traditional irrigation systems include overflow channels that divert excess water during heavy rains, preventing erosion and landslides. During dry periods, water allocation follows strict customary rules that ensure equitable distribution and prevent conflicts that could undermine community resilience.

The Na's livestock management also reflects disaster awareness. Yaks are pastured in high-altitude meadows during summer, avoiding the flood-prone valleys. The practice of maintaining diverse herds—combining yaks, cattle, horses, and smaller animals—ensures that different species can survive various disaster scenarios. Traditional knowledge about animal behavior serves as an early warning system; for instance, unusual restlessness in livestock often precedes earthquakes or severe weather.

Seasonal migration patterns embedded in Na livelihood practices serve dual purposes: maximizing resource utilization and minimizing disaster exposure. The movement between summer and winter pastures, the timing of forest product collection, and the scheduling of trade journeys all consider historical patterns of environmental hazards. This temporal-spatial mobility reduces vulnerability by ensuring that the entire community is rarely exposed to a single catastrophic event.

The collection and processing of medicinal herbs represent another intersection of livelihood and disaster preparedness. The Na maintain extensive pharmacopeias of local plants used to treat injuries and illnesses that commonly follow disasters—infections, respiratory problems from smoke exposure, digestive issues from contaminated water, and trauma. This knowledge is particularly crucial when disasters cut off access to external medical services.

## 3 The Spiritual Dimension of the Na's Disaster Perception

The Na people have been the subject of extensive literature, yet the unique aspect of their culture, the "daba (达巴)" ritual tradition, has not been thoroughly explored. The term "daba" specifically refers to the Na's indigenous ritual specialists and the traditions they uphold. These specialists coexist alongside Tibetan Buddhist monks, known as lama or "dhaba" in Na language, reflecting a diverse spiritual landscape within Na society.

In the Na's rich spiritual universe, encapsulated in the expression "jidzédzin, jimugu," frequently quoted in daba rituals, two fundamental spirit categories are discerned: "dzin (贞)," which occupy the upper realm of the mountains, and "gu (古)," residing in the valleys. These spirits represent the Na's deep connection with their environment, as mountains and valleys are intrinsic to their homeland. The "gu," known as "jimugu," are guardians of clean, drinkable water sources such as wells, fountains, lakes, and seas. Unlike these benevolent spirits, watercourses like rivers and streams are believed to be domains of malevolent entities, polluting the water.

The mountain spirits, "dzin," along with the "gu," are seen as immensely powerful with an ambivalent character, blurring the lines between gods and evil spirits. The Na's celestial gods, "gala," are purely benevolent entities, bestowing blessings and representing the epitome of goodness. However, they do not interact regularly with the ordinary Na. In contrast, the "dzin" and "gu," similar to the malevolent "tsikrua" spirits, are actively involved in the daily lives of the Na people, often causing harm and representing evil forces. This complex spiritual ecosystem shapes the Na's daily life, influencing their rituals, interactions, and understanding of the natural world. These interactions are marked by a deep respect and a careful approach, as the Na navigate their existence alongside these powerful and unpredictable spiritual beings. As a classic shaman formula puts it:

> in the daytime, the dzin do as the gods do
> at night, they act like evil spirits.

The relationship the Na have with their environmental spirits, both "dzin" and "gu," is marked by a profound ambivalence, where these entities can either aid or harm humans based on specific circumstances. This complex dynamic requires the Na to approach these spirits with considerable caution and thoughtfulness. In their rituals, such as the "dzabu" dedicated to the dzin, only designated individuals like the Daba and their assistants are allowed to participate, as these spirits are sensitive to being seen by the general populace.

The "jimugu" or "gu" spirits, sharing the Na's environment, play a significant role in both their social and ritual lives due to their association with essential water sources. The Na's interaction with these spirits is a daily practice, particularly with water collection from wells, known as "dZikrua." These wells, whether natural or constructed, are a central part of village life. The Na also engage with natural springs and lakes, each believed to be under the care of a specific gu spirit. Some of these spirits, referred to as "domestic gu" or "awogu," are even thought to reside within family homes, indicating a deep, ongoing relationship.

The gu spirits are often depicted in a specific form—half woman, half snake. This representation, whether in figurines or drawings, is surrounded by other wild animal figures, particularly snakes and toads, in ritual contexts. According to a Daba song, these domestic gu spirits are also symbolized by everyday objects like the mortar, pestle, and jar, drawing parallels with the shapes of these animals. This rich symbolism reflects the Na's intricate and constant interaction with these spirits, intertwined with their daily lives and beliefs.

> the mill is like the head of the golden frog
> the pestle is like the tail of the golden serpent
> the jar is like the head of the green turtle.

The Na's understanding of atmospheric phenomena like rain and snow is intimately tied to the gu spirits, believed to control water sources and processes like evaporation and precipitation. These spirits embody a deep ambivalence, with the power to either enrich the soil for bountiful harvests or render it barren. This dual nature is encapsulated in a ritual song sung by the Daba, known as the "guémiso" song, which explores the origins and malevolent aspects of the gu spirits. The song poetically describes the devastating effects of droughts caused by these spirits, illustrating their profound impact on the natural world and the lives of the Na people. This aspect of their ritual practice underscores the respect and caution with which they approach these powerful spiritual entities.

> evil spirits gu
> we can't see where they come from
> (they are descendants of) the daughter of the gods in the sky
> who was born of clouds and mists
> (they are descendants of) the daughter of the golden toads on earth
> who was born of the deep seas
> (they) use a grass to dam the water
> the chief of the gu bars the valleys
> he governs all places
> rain does not fall from the sky
> grass doesn't grow on the earth
> toads no longer cry
> cuckoo birds no longer cry
> the tails of lotsorepu birds dry up
> gu medicines no longer exist

Fascinatingly, in Na spiritual traditions, distinct personalities and temperaments are attributed to the dzin and gu spirits. Some of these spirits are known for their demanding nature, often requiring more elaborate offerings and rituals to be appeased. This complexity in their spiritual behavior is particularly evident in the gu spirits, notorious for their ability to ensnare and conceal human souls, an act that is believed to cause illness in the affected individuals. This belief underscores a unique aspect of the Na's interaction with the spirit world, highlighting the need for constant vigilance and intricate ritualistic practices to safeguard against potential harm.

The presence of these spirits, with their dualistic nature, implies a continuous threat to human well-being. However, they are also revered as symbols of fertility and vitality, deeply connected to life-sustaining elements like water sources and rainfall. This duality encapsulates the essence of the Na's spiritual landscape, where the forces of nature are both revered and feared. The Na's ritual practices, therefore, serve a dual purpose: they are a means to cultivate harmonious relationships with these powerful entities and a mechanism to ameliorate any adversarial influences. These rituals, steeped in tradition and belief, form the cornerstone of the Na's approach to coexisting with the powerful and unpredictable forces embodied by the dzin and gu.

The spiritual dimension of disaster perception among the Na provides crucial insights into their disaster response mechanisms. The dzin and gu spirits are not merely abstract religious concepts but active agents in the Na's disaster explanatory framework. Natural disasters are often interpreted as manifestations of spiritual displeasure, requiring both practical and ritual responses to restore cosmic balance.

The ambivalent nature of mountain spirits (dzin) directly correlates with the unpredictable character of mountain-related disasters. Landslides, avalanches, and sudden storms are understood as expressions of dzin anger, often triggered by human transgressions such as unauthorized hunting, excessive logging, or ritual negligence. The Na's elaborate mountain veneration ceremonies (dzabu) serve as preventive measures against such disasters. These rituals, conducted at specific times in the agricultural calendar, aim to maintain harmonious relationships with the dzin, thereby reducing disaster risk.

Water spirits (gu) play an equally significant role in disaster interpretation and management. The distinction between benevolent gu inhabiting clean water sources and malevolent spirits in rivers reflects empirical observations about water safety. Flash floods, contaminated water sources, and drought are all attributed to gu activities. The Na's complex water rituals serve multiple functions: ensuring water quality, predicting water-related disasters, and mobilizing community response when disasters occur.

The guémiso songs performed by daba practitioners contain detailed descriptions of historical disasters attributed to spiritual causes. These oral traditions serve as disaster archives, preserving collective memory of past events and appropriate responses. The songs describe not only the disasters themselves but also the ritual remedies that successfully appeased the spirits and restored normalcy. This knowledge transmission ensures continuity in disaster response strategies across generations.

The concept of soul loss caused by gu spirits provides a culturally specific framework for understanding disaster-related trauma. When individuals experience near-death situations during disasters—nearly drowning in floods, surviving landslides, or experiencing severe weather events—the resulting psychological distress is explained as the gu capturing their souls. The healing rituals performed by daba address both the spiritual and psychological dimensions of disaster recovery.

The spatial organization of ritual activities reflects disaster risk awareness. Ceremonies for mountain spirits are conducted at safe distances from potentially unstable slopes, while water spirit rituals avoid flood-prone areas during dangerous seasons. This ritual geography embodies accumulated knowledge about hazardous zones and safe spaces, effectively mapping disaster risk across the landscape.

## 4 Socio-economic Transformation and Community Resilience

Over the past four decades, spanning from the late 1980s to the present, the Na people have been subjected to numerous external influences. These influences include the processes of globalization, as well as the Chinese state's attempts to assimilate the Na ethnic group with other ethnic groups, such as the Naxi in Yunnan and the Mongols in Sichuan. Additionally, there has been a growing trend of media exposure and mechanization of cultural heritages. Most notably, these external factors have led to significant transformations within Na society and their cultural traditions. These transformations are evident in various aspects of their social structure, kinship systems, household arrangements, economic production methods, and ritual activities.

According to Zhan and al. (1980), in 1956, for instance, there were 1,088 Na individuals residing in 172 households across 11 villages in the Zhongshi area of Ninglang, Yunnan. On average, this equated to 6.33 members per household. In the neighboring Kaiping area, there were 1,057 Na individuals distributed among 174 households, with an average of 6.07 members per household. In some regions, households could be even larger, with reports of households exceeding thirty members in villages in Muli, Sichuan. However, by the late 1980s, substantial changes had occurred. In one Na village, 41% of households had over twenty members, and 66% had over fifteen members, resulting in an average household size of approximately seventeen individuals.

These changes are due, in part, to economic improvements that allowed the Na to renovate their existing houses and construct new ones. However, the new constructions typically accommodate smaller households, reflecting a departure from the larger traditional family units. In a survey conducted in 1979, one Na village in Muli County, Sichuan, had twelve households. Subsequent investigations revealed an increase to thirteen households in 1989, fifteen households in 2000, and twenty-one households in 2013, indicating a significant acceleration in household fragmentation over the past three decades. This trend is evident in other villages in the region as well.

The reasons for dividing households among consanguineous kin can vary, but they still adhere to the social logic of matrifocality within domestic groups. One common incentive for such divisions arises from longstanding disputes between sisters. In such cases, one sister may move out of the household with her children, and sometimes with her brothers, to establish a new residence. Importantly, these disputes have become more frequent due to changes in the Na way of life, where livelihoods no longer solely rely on agriculture. Increasing numbers of Na people migrate to cities like Lijiang and Xichang for work, and others benefit from the region's growing tourism industry. As a result, economic contributions within the same household can vary significantly.

Another reason for household fragmentation is the marriage of a male household member, who must move out to live with his wife. Additionally, there is a trend among the younger generation to get married and move to urban or suburban areas, influenced by popular culture and modern lifestyles.

Historically, the Na people primarily engaged in farming and animal husbandry, with other private economic activities being transformed into collective ownership or suppressed between 1950 and 1980. However, in the 1980s, China's "Reform and Opening" policy led to a diversification of livelihoods among the Na. The first wave of diversification involved transportation, with some households in remote villages participating in the transport and sale of goods like alcohol, cigarettes, tea, snacks, and other consumer products. Others engaged in the trade of livestock, furs, and mountain products. In the 1990s, affluent households in Muli purchased trucks for lumber transport, but this practice was curtailed when the government banned illegal lumbering at the end of the 20th century. During this period, many Na men also ventured into the forests to collect valuable natural products like pine mushrooms and caterpillar fungus, which were sold to Han dealers for profit. Some Na individuals worked as miners in remote areas in Sichuan, Yunnan, or Tibet, while local government construction projects provided short-term employment opportunities.

In the 21st century, the Na people diversified their livelihoods even further. Young men and women found work in tourist areas, working as waitstaff or helpers in locally-run inns and restaurants, and performing traditional dances and songs for tourists. Some migrated to other parts of China or even foreign countries for work or residence.

These economic changes have had a profound impact on the Na people's lifestyles and traditions. The younger generation, influenced by education and urban experiences, develop new tastes in consumption and different ideas about lifestyles, sometimes leading to tensions with their elders and fellow villagers. Traditional agricultural practices are gradually giving way to more modern and urbanized ways of life. As a result, the younger Na generation is increasingly absent from their households, preferring local shops, inns, and video parlors for social gatherings, rather than the traditional practice of gathering in the main house in the evenings.

## 5 The Decline of Daba Tradition and its Impact on Disaster Management

Historical records from the 1920s by American botanist Joseph Rock noted the presence of daba practitioners among the Na, observing their rituals and the growing influence of Tibetan Buddhism in Na society (Rock 1947). However, daba represents a vernacular and more widespread religious tradition among the Na compared to formalized religions. There are no temples or organized structures for the daba, and they do not charge for their services as it is not a profession. Instead, they are regarded as members of households who play an essential role in daily life. Traditionally, daba practitioners are men, reflecting gender-based labor divisions where external relations are primarily handled by men. This also aligns with the apprenticeship system for daba knowledge, which requires close mentoring, and, thus, apprentices learn from masters of the same gender.

During the Cultural Revolution, daba traditions and activities were harshly criticized and suppressed on a large scale. Many rituals were conducted covertly, and even simple ones, like offering food to ancestors before meals, had to be performed discreetly and cautiously. Resource and labor demands made it nearly impossible to organize more significant rituals aimed at expelling malevolent spirits or worshipping deities during that period.

The decline of daba activities during 1960–1980 resulted in a lack of experienced daba masters. Even if apprentices existed, the transmission of knowledge became inconsistent due to social and political pressures. The learning process for daba practices typically spans over a decade, and many apprentices dropped out mid-way due to external pressures. Ritual instruments were also widely destroyed during various social movements, including valuable inherited ones.

The rapid decline in daba activities during 1960–1980 created a situation where the tradition became increasingly rare among the Na. In places like Yongning and Zhuosuo, daba no longer exist, and "half-daba" or apprentices and amateurs who conduct simpler rituals have taken their place. When specific occasions like funerals arise, the Na often have to seek the services of daba from other villages to perform the necessary rituals.

However, some remote Na villages, such as those in Qiansuo and Wujiao in Sichuan, have experienced a revival of the daba tradition to a certain extent. Prior to the 1990s, these villages had minimal road connections to other regions, which limited external influences and administrative interference. Even during the Cultural Revolution, households in these villages continued certain daba rituals privately. Since the 1980s, daba activities have seen a significant resurgence in these areas.

In 1987, Weng Naiqun (Weng 1993) encountered two daba practitioners over fifty years old, as well as four to five young apprentices in a remote village in Wujiao, Sichuan. In another village during the same period, there were seven daba practitioners, with four older than fifty years and three older than thirty years, along with three apprentices following the daba tradition. By 1988, the first village had a population of 205 individuals in 12 households, while the second had a population of 319 in 23 households. These two villages housed the majority of daba practitioners in the entire Na region, encompassing Sichuan and Yunnan. In the late 1990s, three elderly daba passed away in these villages, and in 2012, another daba passed away. Currently, only two daba practitioners over seventy years old remain in these villages, but there are still more than ten daba practitioners in total between the two villages, who continue to serve the Na community and occasionally assist those in neighboring areas.

The lack of experienced daba masters has led to a crisis in the transmission of daba knowledge. As a result, the younger generations, influenced by changes in livelihoods and increased access to education, are less inclined to become apprentices or seek the services of daba practitioners. Some daba practitioners have shifted to other occupations, no longer providing their services or offering limited assistance. Additionally, children and teenagers in these areas now attend schools in neighboring regions or cities rather than following a daba career.

The increasing absence of experienced daba practitioners also has serious implications for disaster resilience. These ritual specialists traditionally served as disaster interpreters, helping communities understand and respond to catastrophic events within culturally meaningful frameworks. They mobilized collective action through ritual, provided psychological support through healing ceremonies, and maintained oral traditions that preserved disaster knowledge. The decline in daba activities thus represents not merely a loss of religious tradition but a weakening of indigenous disaster management systems.

## 6 The Emergence of Environmental Challenges

In the spring of 2004, during my field research in the Qiansuo area of Yanyuan County, Sichuan Province, an unusual incident unfolded. One night, unbeknownst to the local residents, a group of outsiders who were Han people surreptitiously employed high explosives on the Zitawa Mountain in the heart of Qiansuo, aiming to extract ore from its depths. Following the explosion, the region experienced a severe hailstorm, inflicting considerable damage on villagers who were just commencing their spring plowing.

Subsequently, I witnessed the collective outrage of the Na people, the majority ethnic group in Qiansuo. They mobilized a search for the foreign miners and convened at the township government, vehemently demanding legal action by insisting on "going to the county court." Notably, the village chief, upon discovering my affiliation with a university in Beijing, sought my assistance in drafting a legal petition. The written grievances and claims encompassed historical aspects, asserting that during the era of Tusi rule (pre-1956), Zitawa Mountain held sacred significance as a site for the mountain god worshiped by the former Tusi. The villagers contended that they were prohibited from freely ascending the mountain for hunting or woodcutting, let alone detonating explosives. The audacious actions of the outsiders were perceived as provoking the wrath of the mountain god, resulting in the hailstorms that adversely affected innocent villagers. The proposed resolution included identifying and holding the perpetrators accountable for financial compensation, along with organizing a ceremony to appease the angered god.

After navigating several months of complexities, the matter was ultimately brought before the Yanyuan County District Court. The local government intervened, offering reassurances to the villagers, and allocating funds to mitigate the damage inflicted by the hailstorm.

While my personal experience might seem a minor episode in contemporary Chinese rural life, it underscores a crucial reality: at the local level, legal practices and political institutions representing the State often intertwine with indigenous cosmological beliefs, ritual actions, regional ethnic relations, and ecosystems. Furthermore, the actual perceptions, judgments, and resolutions concerning issues of fairness and justice, especially in the case of natural disasters, not only contribute to the creation of a local moral and ethical order but also reflect dynamic shifts in external power relations. As Geertz highlights (Geertz 1983), terms like "haqq" in Arabic, "dharma" in Sanskrit, and "adat" in Malaysian languages are not merely local concepts; they are closer to the Western notion of "right" (Recht, droit) than to "law" (Gesetz, loi). Despite *The Oxford Dictionary of Law*'s definition of law as an institution regulating social relations and behavior through the power of social organizations, anthropological research reveals that legal and political praxis encompasses not only top-down, abstract external rules but also aligns with diverse cultural concepts and corresponding behavioral patterns through extensive fieldwork and ethnographic investigation.

On June 5, 2004, an episode titled "Who is Polluting Lugu Lake" was broadcast on China Central Television's "Common Concerns" program. It shed light on concerns such as the improper disposal of garbage, inadequate wastewater treatment, and unauthorized hotel construction at Lugu Lake. As a result of this broadcast, the ecological issues and pollution problems in Lugu Lake started gaining attention from different segments of society.

Lugu Lake, also referred to as Lugu Hu, is a significant geographical and cultural locale in the region inhabited by the Na. This alpine lake, spanning an area of approximately 50 square kilometers and situated at an altitude of approximately 2,685 meters above sea level, is celebrated for its exceptional natural attributes, including pristine waters and breathtaking scenery. In 2008, Lugu Lake Scenic Area successfully achieved a national 4A-level tourist attraction status. In the year 2020, the Lugu Lake Scenic Area welcomed 280,700 visitors and generated ticket revenue of 8.2099 million yuan. From January to August 2021, the Lugu Lake Scenic Area received a total of 261,800 visitors and generated ticket revenue of 13.6363 million yuan.

The advancement of Lugu Lake's tourism industry has witnessed notable growth in the all-encompassing service sector, encompassing dining, accommodation, leisure, and entertainment. Nevertheless, certain dining establishments and guesthouses have extended beyond the designated boundaries of the scenic area and nature reserves, resulting in the release of contaminants, such as household sewage, directly into the lake. This has raised concerns about the water quality and posed a potential threat to the ecosystem.

Since 2019, the provinces of Sichuan and Yunnan conducted inspections and held symposiums on Lugu Lake to discuss issues related to its protection and governance. They initiated in-depth discussions and established a pioneering cross-province, cross-departmental supervision and emergency coordination mechanism. They also established a high-level inter-provincial lake governance coordination platform. These efforts aimed to coordinate and improve the protection and governance of the entire lake's water environment system across provincial boundaries.

To achieve legal protection and governance of the lake, regional joint legislation was promoted with the coordination of multiple parties. Following approval by the Standing Committee of the Yunnan Provincial People's Congress, the "Lugu Lake Protection Regulations of Lijiang City" were promulgated and implemented on January 1, 2020. This marked the first instance of regional coordinated legislation in the province, addressing governance challenges posed by the "one lake spanning two provinces" and "one lake, two standards" scenarios. It initiated a new phase in the comprehensive protection and governance of the ecosystem, covering mountains, waters, forests, fields, lakes, and grasslands in the Lugu Lake basin. The Yunnan section of Lugu Lake has invested over 1.1 billion yuan in conservation and construction projects. Loshu village, which sits next to Lugu lake, has established a dedicated lake protection team to monitor sewage discharge and residential construction, as well as to carry out river, beach, village, and field cleaning. The number of members in the lake protection team has now increased to over 1,000.

## 7 Conclusion

The study conducted from 2003 to 2019 provides a profound understanding of how the Na people of Southwest China perceive and respond to natural disasters. Central to the Na's perception of natural disasters is their unique cosmology. Their belief system, deeply rooted in ancestral wisdom and nature worship, offers an alternative perspective to the scientific understanding of natural calamities. For the Na, these events are not merely physical phenomena but are deeply symbolic and indicative of larger cosmic imbalances or displeasures of spiritual entities. This view points towards a holistic understanding of disasters, tying them to moral, spiritual, and ecological dimensions.

The Na's livelihood is intricately linked with their environment. Their agricultural practices, community structures, and daily routines are attuned to the rhythms of their natural surroundings. Consequently, natural disasters are not just disruptions of the physical landscape but also profound disturbances to the social and economic fabric of their lives. Understanding this interdependence is crucial for comprehending the full impact of these disasters on the Na community.

The Na's indigenous knowledge, passed down through generations, encompasses a wealth of information on navigating and mitigating the impacts of natural disasters. This knowledge, ranging from traditional agricultural techniques to community-led disaster response strategies, represents a valuable resource for enhancing resilience. It underscores the importance of integrating local wisdom with modern disaster management practices.

The relationship between natural disasters and rituals represents a nuanced intersection of cultural, social, and environmental dimensions, which warrants scholarly investigation within the field of anthropology. Rituals, deeply entrenched in cultural frameworks, hold pivotal significance in communities' responses to disasters. These rituals fulfill various functions, ranging from seeking protection against calamities to restoring societal equilibrium post-crisis. For instance, they may serve as mechanisms for averting malevolent forces believed to precipitate disasters, or as expressions of communal solidarity and resilience amid adversity. Conversely, disasters exert a transformative influence on rituals within communities, often intensifying their significance as coping mechanisms amidst uncertainty and trauma. Rituals offer individuals a structured means of grappling with grief, fear, and loss, thereby imbuing them with a sense of agency in the face of chaos. Furthermore, communities may adapt existing rituals or develop novel ones to address emergent challenges and vulnerabilities precipitated by disasters.

The complex interplay between disaster and rituals is further contingent upon broader socio-economic and political contexts. Cultural norms, religious beliefs, and historical legacies inform the types of rituals enacted and their efficacy in mitigating disaster impacts. Additionally, power differentials within societies shape participation in rituals, as well as the interpretation and valorization thereof.

Anthropological inquiries into the nexus of disaster and rituals yield critical insights into these intricate dynamics, shedding light on the ways in which cultural practices intersect with environmental hazards and social vulnerabilities. By elucidating the role of rituals in disaster preparedness, response, and recovery, scholars contribute to a deeper comprehension of community resilience-building processes. Moreover, such investigations inform culturally sensitive approaches to disaster management, which acknowledge the centrality of local knowledge systems and cultural practices in fostering community well-being and sustainability.

In the face of modernization and globalization, the Na's traditional ways of life, including their approaches to natural disasters, are undergoing rapid transformations. This change presents both challenges and opportunities for the Na. On one hand, there is a risk of losing valuable indigenous knowledge and practices; on the other hand, there is potential for developing new, adaptive strategies that blend traditional wisdom with modern scientific understandings.

The study of the Na's perception of natural disasters, situated within their rich cultural and ecological framework, provides valuable insights into the complex interplay between human communities and their environments. As the world grapples with increasing environmental challenges, the lessons learned from the Na's experiences and wisdom can offer guidance for creating more resilient and sustainable societies. The integration of indigenous knowledge into mainstream disaster management strategies not only honors these traditional practices but also enhances the effectiveness of these strategies, ensuring they are more attuned to the specific needs and contexts of diverse communities.

The insights gleaned from the Na's approach to natural disasters carry significant implications for policymaking, particularly in the realm of disaster management. Policymakers can benefit from incorporating the Na's holistic perspective, which views environmental calamities not just as isolated physical events, but as phenomena deeply interconnected with cultural, spiritual, and ecological aspects. This approach necessitates policies that acknowledge and integrate indigenous knowledge systems alongside scientific methods. Such integration could lead to more culturally sensitive and ecologically sustainable disaster management strategies. Moreover, understanding the Na's interdependent relationship with their environment highlights the need for policies that support and strengthen local ecosystems as a means of enhancing natural disaster resilience. Policies that empower local communities, like the Na, to participate actively in disaster preparedness and response efforts can harness their intimate understanding of local environments, leading to more effective and context-specific disaster mitigation. In essence, the lessons learned from the Na emphasize the importance of viewing disaster management through a multi-dimensional lens, urging policymakers to create frameworks that are not only scientifically sound but also culturally informed and ecologically harmonious.

## Bibliography

Adger, W.N. (2006). Vulnerability. *Global Environmental Change*, 16(3), 268–281.

Bankoff, G. (2004). Time is of the Essence: Disasters, Vulnerability, and History. International *Journal of Mass Emergencies and Disasters*, 22(3), 315–336.

Berkes, F. (2008). *Sacred Ecology* (2nd ed.). Routledge.

Blaikie, P., Cannon, T., Davis, I., & Wisner, B. (2014). *At Risk: Natural Hazards, People's Vulnerability and Disasters* (2nd ed.). Routledge.

Button, G.V., & Schuller, M. (Eds.). (2021). *Contextualizing Disaster*. Berghahn Books.

Cai Hua. (1997). *Une société sans père ni mari. Les Na de Chine*. Paris: Presses Universitaires de France.

Cai Hua. (1999). *Daba—Na shaman* (colour, 40 minutes). Documentary film.

Cajete, G. (1999). *Igniting the Sparkle: An Indigenous Science Education Model*. Skyland Publishing.

Jin Chen. (2012). *Le dualisme na: étude des chants et rituels des Daba (Sichuan et Yunnan, Chine)*. Thèse de doctorat. Paris: École des Hautes Études en Sciences Sociales.

Cutter, S.L., Barnes, L., Berry, M., Burton, C., Evans, E., Tate, E., & Webb, J. (2008). A Place-Based Model for Understanding Community Resilience to Natural Disasters. *Global Environmental Change*, 18(4), 598–606.

Dove, M.R. (2016). *The Anthropology of Climate Change: An Integrated Critical Perspective*. Routledge.

Enarson, E., & Morrow, B.H. (1998). *The Gendered Terrain of Disaster: Through Women's Eyes*. Praeger.

Erikson, K.T. (1976). *Everything in Its Path: Destruction of Community in the Buffalo Creek Flood*. Simon & Schuster.

Gaillard, J.C., Liamzon, C.C., & Villanueva, J.D. (2017). Governance and Community Resilience to Disasters in the Philippines: Lessons Learned from Typhoon Haiyan. *International Journal of Disaster Risk Reduction*, 24, 166–175.

Geertz, C. (1983). *Local Knowledge: Further Essays in Interpretive Anthropology*. New York: Basic Books.

Giddens, A. (1994). Risk and Responsibility. *Modern Law Review*, 57(1), 1–10.

Hewitt, K. (1983). *Interpretations of Calamity from the Viewpoint of Human Ecology*. Boston: Allen and Unwin.

Hirono, M., & Rose, C. (Eds.). (2020). Chinese NGOs at the interface between governmentality and local society: An actor-oriented perspective. *China Information*, 34(3), 279–299.

Hoffman, S.M. (2003). Anthony Oliver-Smith's contributions to the anthropology of disasters: An introduction. *Human Organization*, 62(3), 223–227.

Hsu, Elisabeth. (1998). Moso and Naxi: the House. In M. Oppitz and Elisabeth Hsu (eds.), *Naxi and Moso Ethnography*. Zürich: Volkerkundemuseum.

Huang, Y., Wong, H., & Tan, N.T. (2014). Associations Among Chinese Cultural Beliefs of Adversity, Income Recovery, and Psychological Status of Wenchuan Earthquake Survivors. *Social Work in Mental Health*, 12(4), 343–364.

Huntington, H.P. (2000). Using Traditional Ecological Knowledge in Science: Methods and Applications. *Ecological Applications*, 10(5), 1270–1274.

Jackson, Anthony. (1979). *Na-khi Religion: An Analytical Appraisal of the Na-khi Ritual Texts*. The Hague: Mouton Publishers.

Kidder, Tristram R. and Haiwang Liu. (2017). Bridging Theoretical Gaps in Geoarchaeology: Archaeology, Geoarchaeology and History in the Yellow River Valley, China. *Archaeological and Anthropological Sciences*. 9, 1585–1602.

Le Mentec, K., & Zhang, Q. (2017). Heritagization of disaster ruins and ethnic culture in China: Recovery plans after the 2008 Wenchuan earthquake. *China Information*, 31(3), 349–370.

Mercer, J., Kelman, I., Lloyd, K., & Suchet-Pearson, S. (2010). Reflections on Use of Participatory Research for Disaster Risk Reduction. *Area*, 42(1), 127–137.

Oliver-Smith, A. (1979). On Values, Ethics, and the Conduct of Practice in Applied Anthropology. *Human Organization*, 38(2), 112–120.

Oliver-Smith, A. (1996). Anthropological Research on Hazards and Disasters. *Annual Review of Anthropology*, 25, 303–328.

Oliver-Smith, A. (2019). *Anthropology and Disaster: Reflections on an Evolving Discipline*. Routledge.

Oliver-Smith, A., & Hoffman, S.M. (Eds.). (2002). *The Angry Earth: Disaster in Anthropological Perspective*. Routledge.

Rock, Joseph F. (1947). *The Ancient Na-Khi Kingdom of Southwest China*. Cambridge, Massachusetts: Harvard University Press.

Shih Chuan-Kang. (2010). *Quest for Harmony: The Moso Traditions of Sexual Union & Family Life*. Stanford California: Stanford University Press.

Steward, J. (1955). *Theory of Culture Change: The Methodology of Multilinear Evolution*. University of Illinois Press.

Weng Naiqun. (1993). *The Mother House: The Symbolism and Practice of Gender among the Naze in Southwest China*. Ph.D. dissertation. New York: University of Rochester.

Whyte, K.P., Brewer, T., & Johnson, J.T. (Eds.). (2016). *Value Shift: Why Communities are Key to Solving the Climate Crisis*. New Village Press.

Wisner, B., Gaillard, J.C., & Kelman, I. (2012). *Handbook of Hazards and Disaster Risk Reduction and Management*. Routledge.

Xu, J., Takahashi, M., & Li, W. (2024). Identifying vulnerable populations in urban society: a case study in a flood-prone district of Wuhan, China. *Natural Hazards and Earth System Sciences*, 24(1), 179–197.

Zhang, Q., & Barrios, R.E. (2017). Imagining culture: The politics of culturally sensitive reconstruction and resilience building in post-Wenchuan Earthquake China. In M. Companion & M. Chaiken (Eds.), *Responses to Disasters and Climate Change: Understanding Vulnerability and Fostering Resilience* (pp. 93–102). CRC Press.

陈晋. (2016). "'惹弥' 与' 惹撇': 纳人达巴仪式中的时空认知逻辑", in 民族研究, Vol. 4.

李永祥. (2012).《泥石流灾害的人类学研究:以云南省新平彝族傣族自治县“8.14特大滑坡泥石流”为例》，知识产权出版社。

严汝娴、宋兆麟. (1983).《永宁纳西族的母系制》，云南人民出版社。

云南省编辑组. (1986).《永宁纳西族社会及母系制调查——宁蒗县纳西族家庭婚姻调查之三》，云南人民出版社。

云南省编辑组. (1987).《宁蒗彝族自治县纳西族社会及家庭形态调查——宁蒗县纳西族家庭婚姻调查之一》，云南人民出版社。

云南省编辑组. (1988).《宁蒗彝族自治县永宁纳西族社会及其母系制调查——宁蒗县纳西族家庭婚姻调查之二》，云南人民出版社。

詹承绪、王承权、李近春、刘龙初. (1980).《永宁纳西族的阿注婚姻和母系家庭》，上海人民出版社。

CHAPTER 11

# Disasters, Emotions, and Recovery

*Laurence Roulleau-Berger and Oscar Truong*

## 1 Introduction

In a context of post-disaster in Le Teil, a small French city, we will describe a new humanitarian government (Fassin, 2010); new horizons of coexistence and coordination unfold, and public arenas are opened by institutional actors, intermediate actors, and citizens. Geographies of care and emotional community of *destiny* produce new social solidarities, inequalities, and new moral boundaries; they also sustain new forms of participative democracy; institutional actors and citizens are producing different normative repertories, different competences, and diverse regimes of action (Roulleau-Berger, 2016). Citizens are producing emotional regimes of consent, complaint, indignation, or distancing from the responses provided by the institutions. Interactions between diverse regimes of action are spanning conciliation, negotiation, and disputes. In the post-disaster context, new processes of social differentiation appear, often reinforcing the boundaries between the social groups which existed prior to the disaster. These social differentiation processes play a major role in the production of new inequalities. Meanwhile, individuals are involved in emerging processes of resocialization and in the re-creation of what makes society, notably in *symbolic recovery* (Yama, 2006). Although subjectivities are traumatized, individuals are nevertheless able to mobilize survival, moral, social, political, and economic capabilities. To re-create society citizens are producing collective competences in creating intermediate spaces where the individual fragmented identities could be repaired in a resocialization process.[1]

## 2 Anatomy of a French Small Town: Le Teil

The first historical traces of the municipality of Le Teil date back to the 13th century. At that time, it was only a small village called "Tilliou" that had

1 This chapter is based on the results of the Innovative Program ENS Lyon Anthropocene, inequalities and remaking society in Europe and Asia (2019–2023) led by L. Roulleau-Berger, CNRS/ENS Lyon, Triangle and Y. Yama, Kwansei Gakuin University with O. Truong, G. Giraudo and Pierre Manoury, Triangle- Chen Jin, Tongji University – Zhang Qiaoyun., BNU-HKBU, Beijing.

 | HTTPS://DOI.ORG/10.1163/9789004751804-013

developed around a main castle and was sustained mainly from the cultivation of hemp. During the 17th century, the castle was partly destroyed—only ruins remain today. At the same time, the river trade development brought a second life to Le Teil. Gradually the village expanded, resettling its location on the Rhône river, opposite Montélimar. This position as a port in the middle of a major river trade route transformed the city from a medieval village to a real commercial center. Fabrics and canvases were loaded and unloaded there, as were the many fruits of the Rhône valley. The village of Le Teil was then chosen as a royal salt depot. Unfortunately, the Rhône, a river that is difficult to tame, made it more and more difficult for boats to dock. Flood followed flood, and the river traffic diminished before disappearing completely at the end of the 19th century, to be gradually replaced by rail freight.

Indeed, the Industrial Revolution boosted the activity of the city. Strategically located on the Givors-Grezan line, on the right bank of the Rhône, between Lyon and Nîmes (150 and 110 km away, respectively), the Le Teil train station inaugurated a central depot and hub. It has a major freight vocation but was also an essential transiting platform for local passenger services and goods. Le Teil has been a major hub connecting points on the East–West (Drôme and Ardèche) and North–South (along the Rhône Valley) axes. Back then, the commune sheltered a consequent number of rail workers, up to 1400 for 8000 inhabitants, giving to the city its working-class town features. Many of them joined the Resistance during the Second World War and took part in the "battle of the railways" by sabotaging them. At the end of the conflict, the steam era came to an end, and the city lost its importance as a central hub for the transit of passengers and goods.

Other industries were set up and advanced at the same time the railway was developing (cf Figure 11.1). Among them was the Lafarge cement factory, created in 1848, which quickly became the number one cement company in France. Around 1900 in Le Teil, half of the employed population was working either in the French national railway company or at the Lafarge cement factory. An emblematic working-class housing estate of this era, the Cité Blanche, was the place where most of the Lafarge employees lived at that time. Like most of the workers' housing estates that appeared in the 19th century, it had all the amenities for daily life: dormitories, shops, church. With the decrease of the railway and industrial sectors, the activity of the working-class compound progressively diminished until its closure. Within the same period, the executives of these industries crossed the Rhône to settle in Montélimar. Today, Le Teil, like many small French cities, is suffering from low economic activity, and this impoverished town seeks to regain dynamism.

During the second part of the 20th century, the commune of Le Teil became famous for other reasons. Due to its location between two nuclear power plant

FIGURE 11.1 Rails of trains leading to the former center of the national railway company
SOURCE: OSCAR TRUONG

sites—Cruas, about ten kilometers to the north, and Tricastin, about thirty kilometers to the south—the commune is one of the areas most strongly concerned with the major risk of nuclear disaster. As a riverside town, Le Teil is also likely to suffer from flooding due to its location on the banks of the impetuous river Rhône. Finally, the commune is located in a moderate seismic zone. Nonetheless, what appeared to be the less salient risk threatening this city was ultimately a disaster, as evidenced by the earthquake that occurred on November 11, 2019.

Today, Le Teil is a city of about 8000 inhabitants that is divided into several districts. Among them are two priority neighborhoods (QPVs): the city center and Sud Avenir (composed itself of the two southern districts of Frayol and La Violette). In France, the so-called priority neighborhoods are the identified territories where the city implements a specific policy which aims to compensate the differences in standard of living compared to the national average. With two QPVs, Le Teil is today one of the poorest cities in the department of Ardèche. Regarding the housing situation of those two QPVs, the city center contains a majority of old buildings (built before 1900); Frayol is divided between old and more recent buildings; and La Violette is mainly provided with relatively recent housing constructed during the 1970s. These neighborhoods are therefore those with the highest concentration of low incomes, as well as immigrant populations. In 2016, the share of the immigrant population was 12% for Le Teil, while this figure was 9% for France.

> Le Teil has received all the waves of immigration that France has known. First the Italians, then the Spaniards. The Portuguese arrived much later. Each wave of immigration has experienced racism from the population of Le Teil. My father who comes from Italy told me that in the shops of Teil, one could read "prohibited to dogs and Italians". We blamed the Italians for coming to take the bread of the French and now we hear the same speeches about the new generations of immigrants. At that time, the Teil was a very dynamic city, there was work for everyone, this is no longer the case today. Before, there was the railway station and Lafarge, which employed almost the entire population of Teil. Today, there are only a few people working at the station, few people are still at Lafarge and there is almost no business. (J, retired from National Education at Le Teil)

The two priority neighborhoods (including the city center, Frayol, and La Violette) are the ones where the last generation of immigrant populations is concentrated (cf Figure 11.2). Most of the destruction caused by the earthquake concerns the old habitat, and even more so the old houses and buildings around the earth- quake's epicenter. Close to this epicenter, there are the districts of La Rouvière and Mélas, two neighborhoods made up of old individual houses that were heavily impacted by the earthquake. To these two districts,

FIGURE 11.2 Overview of Le Teil. In the foreground are the city center and its decapitated church. In the background are the southern districts (Sud Avenir) of Frayol and La Violette.
SOURCE: OSCAR TRUONG

where we find mid- dle-class inhabitants of Le Teil, we add a last district in the heights of the city: La Sablière. As its name suggests, the subsoil of this neighborhood is sandy in nature. This district, where we find the wealthiest classes, was affected very little by the earthquake.

Le Teil is a town mainly inhabited by owners of small and modest homes. This is the category of the population that suffered the most from the consequences of the earthquake. The people who rented houses and apartments had the possibility to leave the city, but these small owners even today are still involved in long procedures to rebuild their homes and their lives. Many local actors come into play in this operation: municipality, building workers, social workers, insurance companies, neighbors. Among this disaster-stricken population, the elderly turn out to be particularly vulnerable and concentrate most of the material and moral aid that these local actors are implementing.

## 3 Disaster, Biopolitics, and Humanitarian Government in Le Teil

In Le Teil on November 11, 2019, at 11:52 AM, shortly after the ceremony recognizing the armistice of World War I, an earthquake of the magnitude 5.4 struck the commune. In this low-risk seismic zone it was the most powerful earthquake that had occurred in the region for fifty years. Due to the rarity of such kinds of earthquakes in the region, very few buildings have been built in the city that meet modern anti-seismic standards that are able to handle these types of dramatic episodes. Four people were injured and around a hundred houses and buildings were affected or destroyed. As a precautionary measure, the Cruas nuclear power plant reactors were shut down and only restarted one month later. In the individual stories collected for this research, many people initially believed that the nuclear power plant had exploded. Others had thought there had maybe been an underground mine explosion from the quarry near the Lafarge cement factory, located at the south of Le Teil. Finally, after this short moment of confusion, the announcements made by the firefighters, the municipality, and the local media let the inhabitants understand what had really happened. Among these residents, some who had experienced an earthquake before and elsewhere were able to identify the situation more quickly.

> This day will remain in our heads as the day we never want to see again. We were all at home that day, there was no school, it was a public holiday. So I was at home with my family, I was cooking when everything started to move. I froze at that moment, I was only thinking of my children. I thought it was my gas bottle that had a problem so I quickly turned it

> off. But the tremor continued. The children came to see me, they were scared. I told them to get dressed and we went outside. (N, resident of La Violette)

> I was in the stairwell with a colleague and all of a sudden boom, then a loud noise, then it shook. I thought it was our depot that had exploded, it's not far from where we live, I thought it was a gas bottle or something like that that had blown up our depot. I also wondered if it didn't come from the nuclear power plant, but I quickly learned what had happened with social media. (D, building worker in Le Teil)

The biopolitical framework designed to respond to the emergency situation was first oriented towards securing the victims. Victims of this disaster were called to evacuate their accommodation when necessary and were sheltered in open and requisitioned places, mostly gymnasiums or campsites. Support systems have been added to provide material (housing, food, care) and moral (psychological unit) assistance to these victims.

> We were told to go to this gymnasium, that is what we did. And after, we were asked to go to the other bigger room. And there, really, I can tell you that the first night—I was with my son—the first night, we are reduced to the first part of the Maslow scale. It means it was survival. (X, disaster victim)

The comprehensive reaction that followed the disaster mobilized different levels of aid measures that were quickly implemented following the disaster; they settled down as the basis of a humanitarian government. At the national and regional levels, a state of natural disaster was declared for the affected areas, firefighters from all over the region came to intervene in the city. Experts were also mobilized to establish the first diagnoses, in order to secure the more at-risk populations. At the level of state competency, declaration of a state of natural disaster made it possible not only to release funds to finance the inter- ventions, but also to set up a rehousing unit which has been active for a year. At a local level, the municipality has been at the center of this humanitarian government's action. It is by relying on its expertise and knowledge of local issues that the municipality was first able to direct, organize, and synchronize the interventions of national, regional, and local actors. Following the earthquake of November 11, 2019, a good deal of biopolitical framework was also directly implemented by the municipality to help the affected populations. Simultaneously and in coordination with the municipality's actions, a mobilization of all the civil actors was set up in a spontaneous way. In this surge of solidarity, employees of institutions, members of charities, care professionals, and volunteers have pooled their efforts to set up this humanitarian government.

> How to say ... During the earthquake, the municipality wanted to create a solidarity center. The solidarity pole, it brought together all the solidarity associations that there were on Le Teil. So there was the *Croix-Rouge* (Red Cross), *Arc-en-ciel*, which is an offshoot of *Secours Catholique*. And there was also *Secours Catholique* because there was a person who represented *Secours Catholique* at the national level. Some citizens took care of reception, the Red Cross did what they knew how to do, that is to say everything related to early childhood, to help them, nappies, small pots, the material, the clothes. And then the *Secours Catholique* took care of the people who were in the campsites, and then *Arc-en-ciel* at the reception. And every evening we did a debriefing of what was going on. (PH, Head of Secours Populaire solidarity association, Le Teil)

Once all affected populations had been sheltered, they entered into a dialogue with the assistance systems deployed by the administrative side of the humanitarian government. Despite a situation marked by emergencies, it was over a long period of time that the responses to the needs of the victims were provided. Within this interactional order settled in the aftermath of the disaster, different regimes and relationships emerged according to the solutions provided by the public actors (institutions), the private actors mandated by public actors (associations and experts), and private actors (insurance companies). Interactional orders with the victims in the aftermath of the disaster have been prompted by the nature of the interaction: institutions to victims, mandated actors to victims, and insurance to victims.

If the reaction to the emergency of the situation prompted the humanitarian government to shelter and protect the affected populations, the situation has been handled differently by private actors in charge of compensating those who have suffered material losses. At the personal level indeed, financial assistance for reconstruction depends on each person's insurance. In the years following the earthquake, inequalities in the level of insurance taken out revealed a series of injustices and generated many conflicts between insurers and their customers. In post-disaster situations, the victims are dealing with different grammars of recognition of their experience from the different actors involved in the humanitarian government. While the recognition of their status as disaster victims was symbolically declared just after the disaster by institutional actors, their actual care was provided by actors mandated by the state and local

FIGURE 11.3
Material stored in the "Secours Populaire" warehouse
SOURCE: LAURENCE ROULLEAU-BERGER

institutions. Private actors whose interaction with victims is not mediated by public authorities, however, are part of different registers of recognition. The example of the many post-disaster conflicts between victims and insurers illustrates this situation. It is against these private players, perceived as sinister and lacking in empathy, that victims' anger tends to be directed. Waiting times in an emergency situation, combined with the negotiating relationships in which policyholders find themselves involved, fuel frustration and the feeling of not being recognized as a victim. This lack of recognition clashes with the symbolic recognition granted by the public authorities, creating situations of contradiction within the humanitarian government itself.

In the months following the earthquake, the most affected people were able to benefit from aid and donations from the humanitarian government at the local level. Local institutions and associations created solidarity by offering legal and psychological services and by organizing donation campaigns (Figure 11.3).

At the international level, the humanitarian government has mobilized solidarity from actors who know Le Teil and feel attached to it. This is particularly the case for members of the affected families, or for the twin cities in Germany and Italy that made donations to help with rehousing and reconstruction.

> For example, we have the association of twin towns, a town in Germany whose name I can never remember, they came. They came to visit Le Teil, the Secours Populaire and they made donations to us. We have people who knew Le Teil and who lived in the United States, in Canada, who made donations to us. There, the information went well because people knew about our situation. (PH, Head of Secours Populaire in Le Teil)

## 4 Geography of Care and Emotional Community of Destiny

Disasters, earthquakes, and tsunamis provoke disruptions in normative, social, and moral frameworks. Initially, in emergency situations individuals develop and mobilize collective competences in order to face the catastrophe through relationships of proximity, cooperation, mobilization, and public support (Hasegawa, 2021). The work of aligning commitments ensures the partial re-emergence of the pre-catastrophe normative, social, and moral frameworks, which will appear to be transformed and will produce forms of agreement, alliance, and exchange between different categories of actors that did not necessarily "see eye to eye" before the disaster.

In Le Teil, the mobilizations of the local government, inhabitants, institutional actors, social workers, teachers, and NGOs set up "humanitarian governments" and developed "geographies of care," which signaled an immediate understanding of what was happening, helping and bringing assistance to the victims of the catastrophe. These geographies of care are emerging within the making of competences for individual and collective action and the commitments of institutional actors, doctors, psychologists, social workers, activists, and populations. They are materialized in the organization of assistance, the installation of tents and the opening of gymnasiums for temporary housing, the distribution of water and meals, the implementation of medical assistance, the reestablishment of circuits of communication, information, transport, etc.

> The main reason when people come to us is for food aid. That's the first thing. When we welcome them, we tell them, "Food aid is not a problem, you will get it, but what can we do together?" Our approach is the following: "how can we help you, how can we change your situation." Afterwards, during the earthquake, there were a lot of requests, because there were people who found themselves in the street with nothing, there were structures that were put in place, aid at the material level, furniture, dishes, laundry, etc. It made us change scale, we had a lot of donations to meet these requests. And we continue to relocate people today. When we

> say that an earthquake lasts over time, it's not wrong, we still continue today. These are the aids that appear, it is the visible part of the iceberg.
>
> After, there are plenty of aids: aid for holidays, aid for culture, the cinema, taking them to the theater, taking them to the Mucem[2] in Marseille, things like that ... getting people out of their habits. So, there's culture, cinema, vacations. ... We prepare vacation files for them from February-March so that they can go on vacation for a week in the campsites. That means that they can spread out the financing of their holidays, knowing that on a traditional stay, they will have between 10% and 15% of their holidays insured. And so that's managed and it requires a lot of work, it's more than two months of two full-time. And last year it allowed more than forty families to go on holiday. (PH, Head of Secours Populaire)

In the geographies of care, social hierarchies and positions are suspended, disappearing to the advantage of mutual aid and rescue operations in the aftermath of the earthquake. They reveal how, in a disaster situation, individuals create unusual relationships which spontaneously produce previously unknown forms of mobilization.

> Solidarity has been established between the municipality, institutions, and professionals. Finally, I wanted to talk about the solidarity of people towards people, human solidarity so to speak. And I would like to talk about the meal that took place a week after the earthquake in front of the town hall. I have never seen so many people gathered at Le Teil, everyone was there, everyone cooked something. There was really a need to be together, to come together and show that we were there, all together for each other. There was also a speech by the mayor, it was in front of the town hall. We were really glued to each other and this human warmth made us feel good, it was a very caring moment. (L, habitant of Le Teil)

Progressively the disaster subsides and a local order is constructed, an order in which places are redefined. According to the nature of the disaster, the geographical contexts, the history of the affected places, and professional allegiances, positions are reordered. The systems for the distribution of rehousing and financial aid were organized on the basis of norms of justice.

2 The Mucem is a national museum located in Marseille.

> All the associations, they got together to work together, *Arc-en-ciel* for the sorting, *Secours Populaire* for the stuff, everyone went together to help the victims, to support them. So we were there, at the main school of the city, we took care of the families. My duty was to welcome people and guide them. The family was telling me: "Yeah, I lost that, that, I need that". And I was orienting towards the Poles, that's it. A lot of institutions, associations and services were represented: the Prefecture, the department services, the social landlords, the insurance companies, the lawyers. ... So all these teams went to the same place. As people arrived, they went all the way to get all the information and services they needed to. (A, municipal employee)

But suddenly, these geographies of care were brutally stopped by the outbreak of the pandemic that surged three months after the earthquake in Le Teil, making social boundaries reappear and even reinforcing them. Inequalities between the upper, middle, and working classes, between the *nantis* and the excluded people, became more pronounced during the two following years, especially for those whose homes were destroyed or partially destroyed, and those who were poorly covered by insurance, or who were engaged in pending negotiations to get a more acceptable compensation offer. Facing social, economic, and natural risks brings to light situations of spatial and social injustice that produce an intersectionality of inequalities.

> Many citizens had to leave their homes and Le Teil and felt heartbroken. After the earthquake, we were all in the same situation. But I remember when I went to the emergency unit, I realized the feeling I could have with people who seemed really helpless, who seemed to have lost everything. We felt that some families did not have anything to rely on, that they did not know anyone and that solidarity could not reach them. I was also able to note inequalities about the relation with the procedures and the papers to be completed, some could not do it. I have seen many elderly people who have found themselves accompanied by their children in these procedures. We felt that for some elderly people, it was quite a journey, they had to be rehoused, but they didn't have a car, for example. (H, Social worker in Le Teil)

In this geography of care, the City Council produced a form of participatory democracy in which the grassroots knowledge of the inhabitants and the expert knowledge of architects, urban planners, and academics were brought into dialogue within the framework of a process of production of citizen knowledge to recreate the city and collective life. Elected officials, residents, and actors in the social and cultural sectors insisted on maintaining and even strengthening the social ties that had been built in this emotional community of destiny.

> I am very optimistic, my husband attends meetings with all the people recruited for the reconstruction, there are also citizen consultations with people from the neighborhoods concerned to see how they see the future neighborhood. This earthquake is an opportunity to do great work, to improve what needs to be improved. My only fear is that I will no longer be there to see when everything will be rebuilt. (J, retired from National Education at Le Teil)

The process of emergence of "public arenas" and of different commitment regimes pertains to distinct institutional actors, their forms of mobilization and coordination to produce "interactional citizenship" (Colomy and Brown 1996) and forms of acknowledgment of human suffering. As a result, new and old forms of social coordination are embedded and dissociated in a diversity of institutional and non-institutional places while individual and collective actors are faced with the huge tasks to cope with the ordeal. This takes a central role in the production of moral economies, when, in times of great emergency, caring for others is activated though differentiated forms.

## 5 Moral Economies, Institutional Action, and Inequalities

In the *care economy* we should consider two extreme tendencies: those who are less endowed with economic, social, and symbolic resources tend to receive new moral and material capital, and those who are most endowed with economic, social, and symbolic resources tend to have more agency to deal with the disaster. In this respect, the care economy sustains new forms of social autonomy and domination. The geographies of care thus provoke the construction of moral boundaries between the different categories of institutional actors and the populations which produce situated competences based on the mobilization, selection, and aggregation of resources within differentiated repertories of norms (Swidler, 2001). Moral economies emerge from the group of institutional actors and produce regimes of equity, peace, care for the victims; we may first distinguish different registers of competences among actors of the measures implemented by humanitarian government (Roulleau-Berger, 2017):

- The *technical register*, in which the institutional competence consists in identifying all the needs of the victims and their families.

- The *contractual register*, in which the institutional competence consists in defining and making explicit to the victims their rights concerning rehousing and the indemnification of damage.
- The *civil register*, which concerns the marks of recognition of the ordeal of populations, that is, expressions of courtesy, gestures of support and consolation.
- The *political register*, which is based upon forms of weak, moderate, or strong association of the citizens in the re-creation of society.

The municipality orchestrates the technical register with the experts, the insurance companies, and the state aid linked to the emergency plan (state of natural disaster), which sets up the contractual register, which compensates through the disaster victims' collective. Finally, always supervised by the City Council, the civic and political action are delegated to associations and acts of solidarity coming from the civil society posed, for example, by volunteer psychologists within the help desk organized by associations (housed on municipal premises).

These registers of competence combine and interact to give rise to different models of action subtended by different orders of recognition of the ordeal of the victims founded upon regimes of equity, upon calls for peace, and even upon a weak recognition of the ordeal. The orders of recognition of the ordeal are mainly constructed from conventions and norms stated within the framework of humanitarian government policies for the rehousing and indemnification of victims.

### 5.1 *Recognition of the Ordeal and Regime of Equity*

When the ordeal is recognized in its material and moral dimensions the institutional actors work with a model of action based upon equity. In the register of technical action, they attempt to identify to the best of their ability the needs of the victims and their families. In the contractual register, they define and make explicit to the victims their rights in terms of rehousing and the indemnification of damage. In the civil register, they express marks of recognition of the ordeal of the populations, make gestures of support, of consolation, etc. In the political register, the work of societal refoundation is envisaged as being based upon associations.

> I talk to the people and they are all very happy with the mayor. He was re-elected directly to the last municipal elections (2020), it shows that people are happy with what he has done. So I think the municipality was competent, I'm not in the heart of it, I haven't seen everything but I think they were not bad, they have been very reactive. My clients are above all angry with the insurance companies. (D, building worker in Le Teil)

### 5.2 *Calming the Victims and the Regime of Peace*

When institutional actors use a model of action based upon trickery and calls for social peace, the ordeal of the victims is partially recognized in its material and moral dimensions. In Le Teil measures were set up to produce "registers of peace" which favored the fate of the "good victims," that is, those who accepted that their ordeal could not disturb the life of the local community which had re-formed (Barbot, Dodier, 2010). In the modes of the normalization and supervision of the practices of the victims of earthquakes, the calls for peace imposed upon the victims triggered tensions, complaints, and disputes, revealing inegalitarian modes of redistribution of social, material, and moral resources. They also triggered moral suffering among victims who felt tricked and unrecognized in their ordeal.

> Life is hard, there are ups and downs. To be aggressive, angry, it's useless. But I feel like people here got stuck in their anger. If we adults behave like that, how will the next generation be, what will we tell them? It's very difficult for the children, they don't understand why the comfort they had before has decreased. Let's take an example, before our children had the bus that took them to school. But since the earthquake, there is no more school transport. So we have to take them when we can, but when we can't, they walk to school. When children see their parents who are pessimistic and angry, then they are angry too. We're in a bubble of anger and we can't get out of it. We have to get out of this. (N, resident of La Violette district)

In this situation, the institutional actors do not identify the real needs of the victims and their families; they do not provide them with precise information about their rights to rehousing and getting indemnification; they show them false support and do not consider citizens as having competences for action in the reconstruction of local society.

### 5.3 *Regime of Abandonment and Marginalization*

When institutional actors work using a model of action based upon scorn for the ordeal and upon abandonment, the ordeal is not recognized in its material and moral dimensions. The most vulnerable citizens are marginalized and excluded from the local society.

> About the relocation unit, I found that some people were great, they did their job well, they were very attentive to people's problems. But in the same rehousing cell, some didn't seem very empathic, they aggressively said to the victim "either you go there, otherwise we have nothing for you,"

> I found it was a bit hard. There were victims who asked for solutions for their animals, and these requests were not considered by the relocation services. There were plenty of abandoned cats in Le Teil after the earthquake. Not all the people from the Prefecture who worked in the relocation cell provided similar help and listening, it was a kind of a lottery for the victims who could come across someone good and well prepared or someone worse. I think there were real social workers who were there but also people who were not trained to work with a distressed public. I had the impression that some were bureaucrats, who had ready-made solutions. If people did not fit in the box, there was no solution for them. I found these situations sometimes quite violent. (H, social worker)

The three regimes of action—equity, peace, abandonment of the victims—in the post-disaster context never exist in a pure state or statically. According to the times, places, and political and economic configurations, institutional actors pass from one regime to another and do so according to the competences and regimes of action of the victims.

## 6 Emotional Regimes of Action and Competences of the Victims

Citizens develop economies of political judgment by producing emotional regimes of consent, complaint, indignation, or distancing from the responses provided by the institutions managing the post-disaster situation. The emotional regimes of action of the victims mark out other moral boundaries linked to citizen spaces which arise out of complaint, consent, indignation, or the distancing of institutions.

*Emotional regime of consent* refers to a feeling of belonging to a community of destiny. Citizens are dedicated themselves to help and support the action of the institutions. There is a strong belief and confidence in the competences of the institutions (cf Figure 11.4).

> Because there is still massive social assistance in Le Teil, but it remains calm. We could have people who revolt, people who tell us that after that, after the earthquake, they can't take it anymore, but no! It remains relatively calm and comprehensive. There is a trust. There is great trust in the municipality. There is opposition, that's normal, but I think people are quite grateful to their mayor. And especially since the earthquake

FIGURE 11.4
On the front of the main church of the city located on the main square of the city (in front of the town hall), the municipality posted a call for solidarity and donations.
SOURCE: OSCAR TRUONG

> because it has been very active, it has been very present! We were all present, but he was also a leader. And I think the situation is like this today because of that. (CB, municipality employee)

*Emotional regime of complaint* is brought by citizens who have not recovered from the hardships experienced or are still stuck in administrative procedures. Incomprehension and feelings of frustration regarding deadlines, denial of recognition, or compensation are fueling this regime.

> After the rehabilitation works we also had to lift the evacuation order, but the work was still not finished. Right after the earthquake, I stayed in the house for a week, then the firefighters came to tell me that I had to leave, that the house was too damaged, that it was too dangerous. We were badly affected in the neighborhood (Frayol), we are one of the most affected neighborhoods with Mélas and La Rouvière. It took a long

> time for me to get home, between the insurance, the estimates from the masons. Initially, the insurance only wanted to allocate me 30,000 euros, which is not enough. The expert came back and granted me 10,000 euros more, but I don't know if that will be enough for me. Fortunately, the collective set up by the municipality helped me, because these stories are complicated, and the insurance companies do everything to give us as little as possible. I am at … insurance company, and without the help of the collective, I would have been fooled, I would have had 30,000 euros and no more, I would have had to pay the rest out of my own pocket, I just hope that the 40,000 will be enough, but I don't think so. (S, retired railway worker at Le Teil)

*Emotional regime of indignation* comes as a consecration of the regime of complaint when the person involved in this regime is not heard. They engage individuals in quests for recognition, justice, and reparations to reverse situations of injustice. When these requests are not recognized by the institutions, they can lead to a loss of confidence of the individual towards them.

*Distancing and "the arts of the weak"* (De Certeau, 1980) are the consequence of a breach of trust with the institution. The belief that it cannot help the citizens or cannot meet their demands drives them to develop strategies that take the form of "the arts of the weak." From this moment, citizens will rely on their own skills and networks to try to respond to situations of injustice that the institutions are unable to resolve.

> There were people who were not insured, it was a really difficult situation for them. Afterwards, I don't know why they weren't insured, we went to perform interventions and then we gave the client a voucher so that the insurance paid. At that time, we learned that he wasn't insured. So we did the intervention for free. We had no choice. We put tarpaulins back on the roof to prevent the water from entering the house, we put props so that the house stood up. What can we do when people are uninsured? Poor people, we can't leave them like that. We often do interventions for free, it's not much but we have to show solidarity. Afterwards it is also our pleasure. (D, Building worker)

Here again, the four registers of action of consent, complaint, indignation and distantiation of institutions within a post-disaster context never exist in a pure state or statically within the situations, configurations, and temporalities. According to the times, places, and political and economic configurations, the citizens pass from one regime to another and do so according to the competences and regimes of action of the institutional actors.

## 7 Public Problems and Interactional Citizenship

These moral economies and regimes of action combine using different modes according to the overall contexts and the places and contexts of the situations of conciliation, negotiation, dispute, and exit. Post-disaster situations pose a fundamental question: who is the owner of the social problems which follow disasters? Several answers are constructed around disputed meanings (Gusfield, 2012), answers which are based upon profane knowledge and expert knowledge in the management of disaster-linked social problems. This gives rise to different schematic figures between institutions and citizens within different contexts based upon the regimes of action which combine differently according to the situations of conciliation, negotiation, dispute, and suspicion between institutional actors and citizens.

In the figure of *conciliation*, institutional actors function more within the regime of equity and victims within the mode of complaint or consent. Institutions and citizens are in agreement about the necessity of reciprocal recognition of profane knowledge and expert knowledge in the management of post-disaster life. In this case, institutions and citizens all appear to be owners of the disaster-linked social problems and appear to act together to produce answers.

> It could have happened differently, people could have killed each other. People are still upset since the earthquake and the COVID. When we look at the national news, we can see that people are going crazy just with the COVID, we hear "someone killed someone and so", things like that. It could have happened like that, it's not because we are a small town that things will happen differently. This earthquake kept us together in the end, because if we had only the COVID without the earthquake, it would have been different, maybe people would have been more upset. Everyone at Le Teil experienced the earthquake, it is an experience that we all knew and faced together. (B, Volunteer at Secours Populaire)

In the figure of *negotiation*, the institutional actors subscribe more within the regime of equity and the victims in the mode of complaint, consent, or even indignation. Institutions and citizens debate the legitimization of profane knowledge in the understanding of post-disaster situations as well as expert knowledge which imposes the definition of shared norms. The architects, town planners, social workers, political actors, and inhabitants appear to be partial owners of the disaster-linked social problems and partially act together to find shared answers.

> So we are working a lot on the "Le Teil" of tomorrow. It's almost Le Teil the day after tomorrow because we're over ten years, but it's going to be very gradual. We see it leaving a lot more place for bicycles and pedestrians, that's a first thing, we absolutely want it. We see it trying to be welcoming, hoping that shops will come back, or at least an activity that will come back, it can be crafts, it can be business. Le Teil is a little more human because we want to put in particular in the city center, small parks, things like that so that people have well being, a "Le Teil" where we like to live. We are on it. Then we will try to make an ecological "Le Teil" of course. There are projects, there are many projects, I do not know what will succeed, because it is often very long. There are aids but will they be enough? Afterwards there is a whole urban planning project, we have an assistant who has been appointed to the reconstruction anyway. (CB, municipality employee)

In the figure of *dispute*, the institutional actors subscribe more according to the regime of peace, and the victims in the mode of indignation. In this case, there is conflict around the significations attributed to the disaster-linked social problems. Conflicts arise out of the processes of requesting the legitimization of profane knowledge and the over-legitimization of expert knowledge. The institutions appear to be the owners of the disaster-linked social problems and so the citizens fight to win the case and to appear to be the owners of these social problems.

> The management of this disaster could not have been worse. We no longer know what is happening in Le Teil, the municipality does not inform us. Look for example, on the roof of the bowling alley, the town hall has installed solar panels, and we don't know how much it cost, the energy it produces, where it goes, etc. This is the management of a municipality, we do something for everyone and everyone should benefit from it. It's the capitalist system, it's sad but that's how it is. (P, retired railway worker)

In the figure of *suspicion*, the institutional actors align more according to the regime of scorn, and the victims in the mode of complaint or indignation. On one hand, suspicion is constructed out of the mistrust of the citizens regarding institutions, and on another hand is also based upon the non-consideration and the scorn of institutions regarding victims. The institutions appear to be the owners of the disaster-linked social problems and the citizens appear to be non-owners.

> After the earthquake, the prefect came to see us. She arrived with big statements like: everyone will be relocated. But it was above all a communication operation. It's the same thing with the minister, he came accompanied by journalists just to show that he was there. Both the prefect and the minister arrived with their preconceived ideas and their haughty attitude. It was a way of saying "we are pros and you are amateurs". Yeah, it had that side too. (A, member of the socio-cultural center)

Finally, in the figure of *exit*, which manifests itself in geographical mobilities, the citizens' knowledge about disasters becomes invisible in the population displacements, thus depriving the institutions of bridgeheads from which to revise local and national policies. A total of 15% of the population, traumatized and living in the fear of a new earthquake, left Le Teil in the years following the disaster.

> It seems that there are people who are afraid, yes. There were tenants who left, who may have gone to Montélimar, who may have rebuilt their lives, and then there are people who are saying, "but I won't be coming back to Le Teil, I'm afraid." But now I don't know how to quantify it. Perhaps we should ask the people in the collective. They made a report, very interesting, with 41% of the people who had been affected and who had returned. (CB, municipality employee)

## 8 The Stricken City

In 2022, the daily spectacle of the stricken town for the inhabitants of Le Teil still keeps them in an environment marked by the earthquake. Every day, the sight of the cracks on the walls, the blocked streets, the collapsed or evacuated buildings, the municipal decrees calling on the inhabitants of buildings to evacuate—all these things constantly remind the inhabitants of Le Teil of the earthquake. The slow reconstruction of the city, hampered both by COVID and by insurance procedures, gives the inhabitants of Le Teil the feeling of being stuck in a situation that never ends.

> We arrive at Mélas, so there you see, the whole street has been impacted. There on the left you can see the school, which is soon to be repaired, but there are still the prefabs in the yard. There, on the other side, you can see very old family houses that have been completely devastated. I am attached to this district of Mélas, my father was part of the football team of Mélas, that's why it hurts my heart to see Mélas like that. In La Rouvière, a little higher up, I know people who were born in houses that are now completely destroyed. Look in this street, you can find scars from the earthquake on all the houses. Every day when we walk in the city, we see things that remind us of the earthquake. (J, retired from national education at Le Teil)

In the aftermath of a disaster, the stricken city is impacted by the disorganization caused by the earthquake (cf Figure 11.5). Because of the population displacements it generates, neighborhood relations are initially weakened, the frequentation of places is modified by destruction, which pushes municipal services or associ- ations to move out from their collapsed spaces. On the front line in the direct aid provided following the earthquake, the relocation of these

FIGURE 11.5
Collapsed building at Le Teil
SOURCE: OSCAR TRUONG

services raises confusion for the inhabitants of Le Teil seeking help. In terms of municipal or associative infrastructures' capacity to respond to requests, the massive influx of disaster victims creates situations of overload for the services, which are quickly overwhelmed by the emergency. The humanitarian government that was established had to carry out the work of readjusting the geographies of care in order to carry out its actions.

> We requisitioned the school in the center, because it was also affected. So one part was not damaged and we requisitioned the part that was not damaged, that's where all our offices were. Finally we received the victims there. And like that, we helped the town hall. Everyone was arriving at the town hall. There were queues going all the way here, we didn't know what to do. And the worst part of all that is that we inherited people coming from other cities, Viviers ... the victims of Viviers, the victims of Rochemaure. (A, communication officer at the municipality)

There are many traumas following this event. Identities were first weakened by the trauma of the experience of the earthquake. Many residents of Le Teil have, in the months following this event, apprehensions about the possibility that the earth will shake again. Each suspicious noise or unusual event brings people back into the memory of a trauma and the fear that this disaster will repeat itself.

> Yes, so it's the stomach ache the first nights when you sleep in the house. So the first few days, we slept on the ground floor in the living room because we didn't dare go into the bedrooms. Because everything is marked, everything is cracked, the house is very marked! And then afterwards, we go to bed with fear in our stomachs. And even now, so I go to bed with more fear in my stomach, but there must be no noise. When there is wind. When it's windy I know I'm really, really scared. Because we still have a very, very fragile chimney on the roof. (CB, municipality employee)

The identities of certain inhabitants of Le Teil are, like their towns, stricken. When they do not have the will or the possibility to leave the city, some affected inhabitants have been in temporary rehousing situations for years waiting for the reconstruction of their housing.

> People were more traumatized at the beginning more than now. During the period just after the earthquake, at the moment of the intervention of

> experts, insurance, all that, it was a really hard time for the victims. There are people who only received the money the day before yesterday, a year and a half after the earthquake. I have noticed that people start getting better the moment their house is rebuilt. I have a client, Mr. ... who was completely depressed, his wall was completely gone, his house was half destroyed. We built a new one, we rebuilt his roof, the interior a bit, the facades, and as soon as we worked at his place, the smile returned. It's the waiting that depresses, people rebuild themselves at the same time as their houses. The hardest thing is for the old ones, if you take away their daily life, their habits, there is nothing left. I have a friend in La Rouvière, he was at home and there, brrrrouuuaaah, no more house, we had to raze it, and it was a big building too. In six seconds of earthquake, some have lost everything, and they cannot move on until they are resettled in their homes or somewhere. (D, Building worker)

## 9 Human Recovery and *Intermediate Spaces*

In post-disaster contexts, although individual subjectivities are traumatized, individuals can nevertheless be able to assemble individual and collective capabilities of social and economic organization. New horizons of coexistence and coordination unfold and public arenas are opened by institutional actors, intermediate actors, and citizens. They are all involved in emerging processes of re-socialization and the re-creation of what makes society (Nagata, 2016).

Rebuilding society in this impoverished and disaster-stricken city is being organized with the emergence of intermediate spaces. For example, the "Tiers Lieu" space in Le Teil, created in 2019 by a collective of activists and social workers, can be defined as an intermediate space of social recomposition (Roulleau-Berger, 1999, 2003) based on symbolic recovery (Yama, 2006, 2022). It is a permaculture space but also a place that contains coworking, a shared workshop, a fablab, a garage, a maker space a multimedia and digital space. This place is designed as a new place of social linkage, emancipation, and collective initiatives. Since 2019, several associations, collectives, self-entrepreneurs, companies, volunteers, craftsmen, and local producers have wanted to share their projects and find together a new permaculture and socio-cultural synergy in Le Teil (cf Figure 11.6). The idea today is to collectively create a coworking space, a room for welcoming and sharing projects, shared workshops (dance, theater, circus, yoga) with qualified trainers, a workshop for creation and scenography, a media center for information and entertainment, as well as a reception area for all (schools, colleges, etc.). During the COVID-19 pandemic

FIGURE 11.6
Tiers Lieu Le Teil
SOURCE: LAURENCE ROULLEAU-BERGER

and after the earthquake, the birth and life of this heterogeneous place concentrated social creativity, reinvention of ecological lifestyles (biometal, waste recovery, etc.), and cultural activities that require forms of political and social autonomy. This collaborative space proposes new forms of democratic life on a small scale that produces autonomous public spaces (Habermas, 1992) in a fragmented, dis- aster-stricken but resilient city, capable of rebuilding social ties and allowing a local community to emerge, both ecological and cultural, of recovery and reinvention of the social.

The theater company La Ligne has set up a form of participatory theater from the staging of stories of disaster by the inhabitants of Le Teil. We could speak here of an *intermediate space of cultural recomposition* (Roulleau-Berger, 1999, 2003).

> In this cultural project that today we call "Identity of Le Teil," maybe the name will change, I don't know, at the beginning when we made it there were indeed these three themes: exhibition, the multimedia course and thus a project of play with the company *La Ligne*. So that was well advanced. We are still in it, except that the project of *La Ligne*, which was basically to work on the questions of migration, work, and the migra- tory path of the inhabitants, has necessarily evolved. There was the earth- quake and in fact, they ... they thought about it and they started from another point of departure. They took the earthquake as a starting point to make people talk, because we felt that people needed to talk too, to let go a little. So ... well, they will tell you more about it, but their project of a participative theater piece has evolved a lot because of

> the earthquake, so … so there you go. So for the last year, they have been working on it, they have … they have done a lot of things. They have done a lot of collect- ing of words from the inhabitants, especially just after the confinement, it seems to me, in June-July, in various places, in the market, in different places of the city, in associations, so there you go. They were able to … they had great testimonies and as a result, they try to create groups like that of inhabitants who are going to continue the project, without obligation to go until the end, but there you go. And there, at the moment, they are on … on the animation of workshop with some inhabitants, to really think about this play. The initial calendar foresees a play for February–March, but well, with … with the COVID we don't really know if we will be able to maintain this calendar, but … but there you go. Well, you will talk about it with them later, because indeed, the earthquake has really, I think, had an impact on their project of a play. (AD, in charge of communication and citizen participation at the municipality)

Risk societies produce *intermediate spaces* in which collective and individual actors reappropriate resources in an attempt to develop societal capabilities. In a context of diverse uncertainties, intermediate spaces have both become more diversified in form and hierarchized on the basis of modes of socialization. As the situation of citizens has become increasingly precarious, institutions have declined, social and spatial inequalities have increased, and discrimination against ethnic groups has intensified, so intermediate spaces have proliferated, reflecting forms of citizenship characterized by defiance and protest. In Le Teil they can be described as 'moral regions' where the vulnerable social groups mobilize, after the earthquake, create their own strategies for survival, and set up micro-organisations of an economic, cultural, or artistic nature.

These intermediate spaces occupy positions alongside spaces set up to produce discreet, liminal forms of socialisation that are constructed in non-institutional worlds. Within these intermediate spaces, populations mobilize to produce material and symbolic economies and delineate the boundaries of spaces in which they feel recognized. This in turn gives rise to *cultures of uncertainty*, which are based on the management of double-bind situations and the construction of collective identities and may give individuals access to recognition by others. These cultures give rise to small-scale social practices taking one or more forms that reveal individuals and groups are also capable of invention in a context of post-disaster, that is, able to take action and endowed with societal capabilities. The intermediate spaces contain the discreet, even invisible and discrete social forms characteristic of the dispersed, tactical, and makeshift creativity of citizens, the 'art of making do' and the tactics adopted by the 'weakest' against 'the strongest' (De Certeau, 1980). Some of the individuals who are denied certain forms of social recognition develop forms of collective resistance by creating these intermediate spaces. They bring to light a social struggle that takes the form of a practical process in the course of which individual experiences of 'distancing' and contempt are collectively reinterpreted. The intermediate spaces are formed out of situations in which entire populations are 'distanced'; these situations may in turn give rise to conflicts, social movements, and riots. However, while they may regenerate social affiliations, they may also produce new forms of social exclusion.

In this context, intermediate spaces take place in existing or new associative structures. Former bonds of solidarity are held and expressed by a population of retired people as an inheritance of Le Teil workers' memory.

> It's this link between history, migration, work, rebuilding a society after the earthquake. Finally here we are in a dimension where we are also thinking about reconstruction, urban renewal. ... And all that in fact is ... here, renewal, reconstruction is a change. And change is scary. Within that, in fact, we also have to integrate the inhabitants in this entire process in terms of creation, to involve them, it is also to try to reassure them about the future of the city, about their future as inhabitants. As a municipality employee, it is our will, really. That's it, we want to show them that the city itself moves all the time, it changes all the time, and that's part of its history. What is good is to be able to get involved in this change too. (A, communication officer at the municipality)

## 10 Conclusion

From the perspective of the social embedding of natural disaster in Le Teil, the disaster is more or less integrated and socialized as an expected event within the course of collective and/or individual life. Individuals' psychological lives and subjectivities are partially or totally fragmented. Continuities between their lives before and after the disaster could produce gaps within identities, but individuals are also able to develop capabilities. For social groups and individuals the question of the production of a collective memory also becomes vital in the process of reconstruction of individual and collective identities. We are coming back to a fundamental sociological question: What makes society?

While situations of ecological disasters produce paroxysmal patterns of fragmentation, rupture, and physical and social destruction, they also open up spaces for social restoration and the reconstruction of societies. The challenge is still to understand how the voices of citizens are heard publicly or when disasters are over, and how collective, individual, and even institutional memories participate in the process of re-creating societies and emotional communities. If new horizons of coordination and new public spaces are produced by institutional actors, intermediate actors, and citizens to "rebuild society," new situations of societal inequalities and social injustices, new moral boundaries between different social groups in a post-disaster context, are formed.

## Bibliography

Barbot, J., Dodier, N. 2010. "Violence et démocratie au sein d'un collectif de victimes. Les rigueurs de l'entraide." Genèses, 2010/4, n°81, pp. 84–103.

Certeau, M. De. 1980. *L'invention du quotidien, I: Arts de faire*, Paris: Gallimard.

Colomy, P.J., Brown, D. 1996. "Goffman and Interactional Citizenship." *Sociological Perspectives* 39 (3) (octobre 1): 371–381.

Fassin, D., 2010. *La raison humanitaire*, Paris : Seuil

Gusfield, J. 2012. "Significations disputées. Contester la propriété et l'autorité des problèmes sociaux," in D. Cefaï and C. Terzi: L'expérience des problèmes publics, Editions de l'EHESS, Paris, pp. 113–135.

Habermas, J. 1992. "L'espace public", 30 ans après, *Quaderni*, n°18, pp. 161–191.

Hasegawa, K. 2021. "Japanese Environmental Sociology: Focus and Issues in Three stages of development." International Sociology Reviews 36 (2) 289–301.

Nagata, M. 2016. "A bottom-up counterpart assistance approach for the revitalization of communities after an extensive disaster," in Li Peilin and Laurence Roulleau-Berger (eds.). *Ecological Risks and Disasters: New Experiences in China and in Europe*, London & New York: Routledge, pp. 157–170.

Roulleau-Berger, L. 1999. *Le travail en friche. Les mondes de la petite production urbaine.* La Tour d'Aigues: Editions de l'Aube.

Roulleau-Berger, L. 2003. "Economie solidaire, non-reconnaissance publique et production d'espaces intermédiaires," in *Economie solidaire et démocratie, Revue Hermès*, deuxième trimestre.

Roulleau-Berger, L. 2016. "Life exposed, inequalities and moral economies in post-disaster societies: China, Japan and Indonesia." In Li Peilin and Laurence Roulleau-Berger (eds). *Ecological Risks and Disasters: New Experiences in China and in Europe*, London & New York: Routledge pp. 135–155.

Roulleau-Berger, L. 2017. "In commemoration of the legacy of Ulrich Beck: theory of migration and methodological cosmopolitanism." *Development and society*, septembre 2017, vol. 46, nº 2, pp. 227–250.

Swidler, A. 2001. *Meaning and Modernity: Religion, Polity, Self* (University of California Press, 2001).

Yama, Y. 2006. 「『象徴的復興』とは何か」 [What is Symbolic Recovery?]. *Sentan Shakai Kenkyu*, 5.

Yama, Y. 2022. 「語り合う場のデザイン一哲学カフェの試みから」 [Designing Dialogical Spaces: The Philosophy Café Approach]. *DPRI Newsletter*, 101, 7. Kyoto: Kyoto University Disaster Prevention Research Institute.

# Index



 | HTTPS://DOI.ORG/10.1163/9789004751804-014